Semiconductor Raman Lasers

For a complete listing of the *Artech House Optoelectronics Library*, turn to the back of this book.

Semiconductor Raman Lasers

Ken Suto
Jun-ichi Nishizawa

Artech House
Boston • London

Library of Congress Cataloging-in-Publication Data
Suto, Ken
Semiconductor Raman lasers / Ken Suto, Jun-ichi Nishizawa
Includes bibliographical references and index.
ISBN 0-89006-667-1
1. Semiconductor lasers. 2. Laser communication systems. 3. Raman effect. 4. Brillouin scattering.
I. Nishizawa, Jun-ichi. II. Title.
TA1700.S88 1994 94-21067
621.36'6–dc20 CIP

British Library Cataloguing in Publication Data
Suto, Ken
Semiconductor Raman Lasers
I. Title II. Nishizawa, Jun-ichi
621.366

ISBN 0-89006-667-1

685 Canton Street
Norwood, MA 02062

International Standard Book Number: 0-89006-667-1
Library of Congress Catalog Card Number: 94-21067

10 9 8 7 6 5 4 3 2 1

Contents

Preface

Stimulated Raman scattering is a well-known nonlinear optical phenomenon, which has generally been thought to require very high pump power. In the early 1960s, J. Nishizawa proposed that stimulated Raman scattering in semiconductors could be used for frequency conversion and other techniques in optical communication, although practical optical communication did not exist at that time. Raman laser oscillation in a semiconductor was first observed in 1980. Since then, we have been developing low-threshold semiconductor Raman lasers based on a GaP—$Al_xGa_{1-x}P$ heterostructure waveguide. Optical interaction in a waveguide is so effective that the threshold pump power has been drastically reduced.

The semiconductor Raman laser has unique functions that are applicable to optical communication and other applications and cannot be realized by any other kinds of lasers. One of these features is a frequency selective lght amplification, which is applicable to heterodyne-type demodulation of wideband-modulated light signals.

The book is devoted to the fundamentals and practical technologies for the semiconductor Raman laser. This includes comparative discussions of rare-earth doped optical fiber amplifiers and laser diode amplifiers. These discussions will give a perspective view for future very wideband optical communication, which can be called terahertz-band optical communication.

The authors would like to express their appreciation to T. Kimura for cooperation through the development of the semiconductor Raman laser. They are indebted to Research Development Corporation of Japan for the support provided as a part of Nishizawa Terahertz Project in ERATO system. The authors also express their thanks to their colleagues for assistance in handling manuscripts.

Chapter 1
Introduction

1.1 SEMICONDUCTOR RAMAN LASERS FOR OPTICAL COMMUNICATIONS

Optical communications currently use only one ten-thousandth part of the total information capacity of a light wave. From the beginning when Nishizawa proposed the semiconductor injection laser [1,2], the semiconductor Raman laser and the Brillouin laser were considered to be as well suited for truly wideband optical communications [3,4]. It was expected that light waves could be handled just like electromagnetic waves in conventional electrical communications, including frequency conversion and heterodyne demodulation.

The frequency at which amplification occurs in a Raman laser and in a Brillouin laser is not fixed but is dependent on the pump laser frequency. Therefore, the essential function of the semiconductor Raman laser is frequency-selective light amplification, which cannot be achieved with other kinds of lasers. Thus, as presented below, the most important application of the semiconductor Raman laser is heterodyne-type demodulation in a wideband optical communication system.

The fast and efficient light detectors, the pin photodiode and the avalanche photodiode, which are mainly used in the present optical communication system, were proposed in very early years [5–7]. Schottky diodes also show fast response. Although they have excellent characteristics, their response frequencies are far below the terahertz frequency region. Therefore, heterodyne-type optical demodulation should be a key technology in a future terahertz-band optical communication.

Second, there is a possibility that the semiconductor Raman laser can be used as a light source with much higher coherency than a laser diode. This is because the resonator of a semiconductor Raman laser has a very high Q-factor.

Third, far-infrared coherent radiation can be generated in a semiconductor Raman laser by using a polariton mode or by pumping with two light beams of different frequencies. In spite of all these excellent functions, studies of Raman lasers were rather limited because they required high pump powers on the order of a megawatt. However, we have shown that semiconductor Raman lasers having a waveguide structure with a GaP core and $Al_xGa_{1-x}P$ cladding layers can be operated with a very low pump power, much less than 1W, which will make the semiconductor Raman laser a very practical device.

This book first describes the history of various kinds of Raman lasers and Brillouin lasers. It also discusses the present situation and the future of optical communication, in which Raman lasers are thought to have an important role. Chapter 2 discusses the fundamentals of stimulated Raman scattering and the Raman amplification mechanism in semiconductors as well as in glass-fibers. Then, the bulk GaP Raman laser and the GaP-$Al_xGa_{1-x}P$ heterostructure waveguide Raman lasers are described in detail, which will make clear the essential properties of the semiconductor Raman laser and the importance of waveguiding for low-threshold operation. Fundamental fabrication techniques such as high-quality epitaxial growth of GaP-$Al_xGa_{1-x}P$ layers and reactive ion etching (RIE) techniques for waveguide formation are described in Chapter 6. Chapter 7 makes a comparative study of the light amplifiers that are thought to be key devices in future optical communication systems: the erbium-doped fiber amplifier, the fiber Raman amplifier, the laser diode amplifier, and the semiconductor Raman laser amplifier. In addition, the Brillouin laser is discussed in Chapter 4. Finally, Chapter 8 addresses the highly coherent properties of the semiconductor Raman laser oscillation and the ability of efficient frequency mixing in the semiconductor Raman laser.

1.2 HISTORY OF STUDIES OF RAMAN AND BRILLOUIN LASERS

In 1962, Woodbury and Ng first observed a strong Raman-shifted emission line, at $\lambda = 7{,}658\text{Å}$, appearing from the nitrobenzene Kerr cell that was used for *Q*-switching of a ruby laser with $\lambda = 6{,}943\text{Å}$ [8]. This was the first observation of stimulated Raman scattering. Later, various gas, liquid, and solid materials were found to show stimulated Raman scattering when they were pumped with strong laser light, as shown in Table 1.1 [9].

As illustrated in Figure 1.1 [23], not only the first order Stokes line with frequency $\omega_{s1} = \omega_L - \omega_{ph}$, but several higher order Stokes lines $\omega_{sm} = \omega_L - m\omega_{ph}$, as well as anti-Stokes lines $\omega_{Am} = \omega_L + m\omega_{ph}$, were observed, where ω_L is the pump laser frequency, and ω_{ph} is the optical phonon or vibrational frequency.

Various effects related to the stimulated Raman scattering were interpreted in terms of the induced field via lattice vibrations by Garmir, Pandarase, and Townes [24]. Bloembergen and Shen developed the theory of optical nonlinearity based on the concept of the nonlinear susceptibility, which greatly contributed to the general understanding of the optical nonlinear effects including stimulated Raman scattering [23,25].

Table 1.1
Typical Materials for Which Stimulated Raman Scatterings Were Observed*

Material	ν_R *(cm^{-1})*†	$\delta\nu_R$ *(cm^{-1})*†	$Nd\sigma/d\Omega$ *($10^{-6}m^{-1}str^{-1}$)*	g *($10^{-5}m/MW$)*	*Reference*
H_2 (100 atm)	4,155	0.20	3.0×10^3	1.5	[10]
Water	3,420	176	0.51	0.07	[11]
Acetone	2,921	18	5.4	1.2	[12]
Cyclohexane	2,852	11	3.7	1.2	[12]
N_2 (liquid)	2,326.5	0.067	0.29	17.5	[13]
O_2 (liquid)	1,552	0.117	0.48	14.5	[13]
Nitrobenzene	1,345	6.6	6.4	2.1	[14,16]
Toluene	1,003	1.94	1.1	1.2	[15,17]
Chlorobenzene	1,002	1.6	1.5	1.9	[14]
Bromobenzene	1,000	1.9	.5	1.5	[14]
Benzene	922	2.15	3.06	2.8	[13,14]
CS_2	655.6	0.5	7.55	23.8	[13,17]
CCl_4	459	5	2.3	0.85	[12]
Diamond	1,332	2.04	17	6.9	[18]
Calcite	1,086	1.1	2.9	4.4	[19]
$BaNaNb_5O_{15}$	650			6.7	[20]
	655			18.9	
Silicon	521	0.8	305	190	[21]
Quartz	467	6.7	3.1	0.8	[20]
Li^7TaO_3	201	22	238	4.4	[20]
	215	12	167	10	
Li^6TaO_3	600			4.3	[20]
	608			7.9	
Li^7NbO_3	256	23	381	8.9	[20]
	258	7	262	28.7	
	637	20	231	9.4	
	643	16	231	12.6	
Li^6NbO_3	256			17.8	[20]
	266			35.6	
	637			9.4	
	643			12.6	
InSb ($n = 3 \times 10^{22}m^{-3}$)	100	2	30	16.7	[22]

*After [9].
†1 cm^{-1} = 30 GHz.

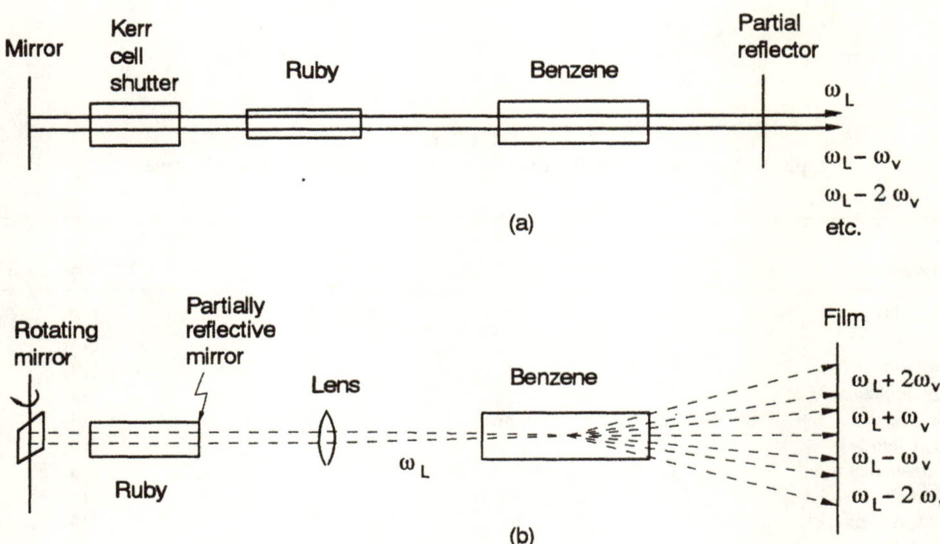

Figure 1.1 Experimental arrangements: (a) experimental arrangement of a Raman laser, and (b) experimental arrangement for the production of stimulated Stokes and anti-Stokes light in an externally focused laser beam [23].

In the early years, experiments on stimulated Raman scattering required very high pump powers of the megawatt or even gigawatt order. These powers were required not only because Raman polarizabilities of most of the materials being studied were rather small but also because the only method for optical interaction was to focus a pump light beam on a bulk medium by using a lens.

There was another direction in the studies of the Raman lasers. Nishizawa aimed at optical communication when he invented the semiconductor injection laser in 1957 [1,2]. He proposed that lattice vibrations in semiconductors could be used for frequency conversion to generate far-infrared coherent light and other techniques that would become essential in the optical communication systems he suggested [3,4]. Later, the semiconductor Raman laser was implemented in 1980 using a GaP crystal by Nishizawa and Suto [26]. Although the first semiconductor Raman lasers needed a pump power close to 1 MW, the pump power was reduced to less than 1 W by introducing a waveguide structure composed of GaP-$Al_xGa_{1-x}P$ heteroepitraxial layers [27]. This was considered a practical power level for applications in optical communications.

A great advantage of semiconductors is that appropriate waveguiding of the Stokes light as well as the pump light can be designed on the basis of heterostructures. Figure 1.2 illustrates the semiconductor waveguide Raman laser.

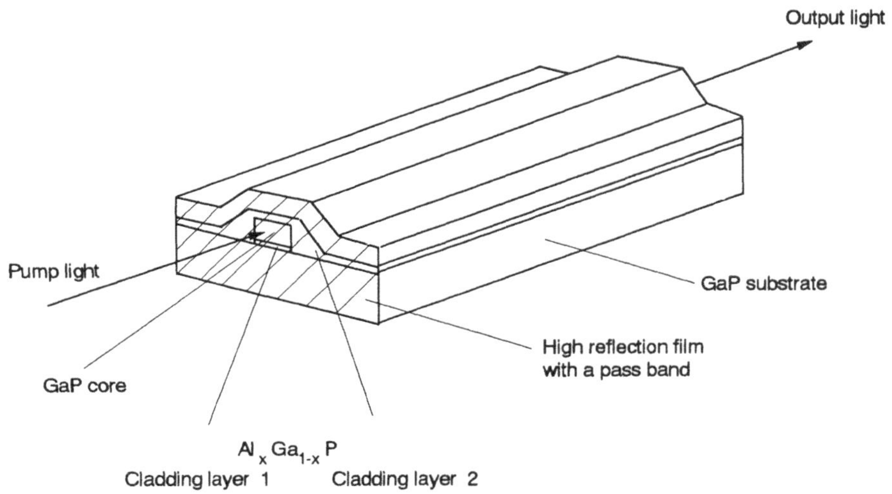

Figure 1.2 Semiconductor waveguide Raman laser.

Another advantage of semiconductors is that the Raman polarizabilities are thought to be high because resonance enhancement occurs when the direct bandgap energy is close to the photon energy of a pump light.

So far, we have discussed stimulated Raman scattering and Raman lasers related to lattice vibrations. There is, however, another kind of Raman laser based on spin levels of the conduction electrons in semiconductors, which is called the spin-flip Raman laser. In 1970, Patel and Shaw realized a spin-flip Raman laser in an InSb crystal pumped by a CO_2 laser at a wavelength of 10.6 μm [28]. The pump power level was reduced to the level of several watts. However, the spin-flip Raman laser has the disadvantage of requiring a low temperature. It is interesting to note that a frequency conversion experiment was performed by the Lax group using spin-flip Raman lasers [29].

Among various solid-state materials, $LiNbO_3$ and related crystals were extensively studied because they have large values of Raman polarizabilities [30]. Yarborough and others showed that generation of continuously tunable far-infrared coherent radiation in the wavelength range from 238 to 50 μm was possible with Raman scattering via polariton-mode optical phonons in $LiNbO_3$ [31].

On the other hand, Stolen, Ippen, and Tynes reported Raman laser oscillations in an optical glass fiber [32]. Although the silica glass itself has an extremely small value of Raman polarizability, the Raman laser oscillation occurred as a result of propagation in a long fiber (several meters in length). The pump power was several watts or less. When the fiber Raman laser is used as an amplifier that needs no resonator structure, the fiber length can be extended for as long as several tens of kilometers, with the pump power reduced to several tens of milliwatts. This fact indicates the importance of waveguiding in obtaining effective interactions between light waves.

Practical applications of the Raman lasers were limited so far. An H_2 gas Raman laser was used for wavelength shifting from a short-wavelength, high-power laser such as pulsed dye lasers. However, low-pump-power waveguide Raman lasers, like the fiber Raman laser and the semiconductor waveguide Raman laser, will greatly extend the field of application.

Instead of the molecular vibrations or optical phonons involved in stimulated Raman scattering, acoustic waves are involved in stimulated Brillouin scattering. Chiao, Townes, and Stoicheff excited a quartz crystal by focussing a giant pulse beam from a ruby laser and observed a strong Brillouin shifted emission line with a frequency shift of nearly 30 GHz in the backward direction [33].

Table 1.2 shows various materials in which stimulated Brillouin scattering was observed [9].

In early experiments, very high pump powers were necessary, similar to the stimulated Raman scattering experiments. However, Ippen and Stolen observed that for the backward stimulated Brillouin scattering from a silica optical fiber, the threshold pump power was less than 1 W if a very coherent pump source equipped with

Table 1.2
Typical Materials for Which Stimulated Brillouin Scatterings Were Observed*

Material	ν_R *(MHz)*	$\delta\nu_R$ *(MHz)*	γ	g *(10^{-5}m/MW)*	*Reference*
N_2 (120 atm)	1,170	20	0.075	23	[34]
Acetone	4,600	180	0.99	19	[35–37]
Ethyl alcohol	4,550	350	1.09	12	[35]
Ethyl ether	3,900	185	0.99	24	[38]
CCl_4	4,390	650	1.35	6	[37,39]
Cyclohexane	5,550	670	1.37	6.8	[35,39]
Toluene	5,910	480	1.73	13	[37,39]
Nitrobenzene	6,840	900	2.06	7	[34]
CS_2	5,850	75	2.37	128	[37]
n-hexane	4,290	220	1.07	26	[37]
Benzene	6,470	320	1.62	18	[38,39]
Water	5,690	380	0.87	4.8	[38,39]
Methyl alcohol	4,250	230	0.91	13	[37,39]
Glass (BK)	22,200	80	0.22	0.7	[37]
Quartz	25,500	30	1.41	6	[37]
Fused quartz	24,300	55	1.22	6.6	[40]
Lucite	12,000	106	2.41	20	[37]

*After [9].

an internal etalon was used [41]. It was shown that the bandwidth for the amplifying process was much narrower than that for Raman processes, usually narrower even than the linewidth of conventional pump lasers [42]. As a result, only a part of the pump power contributed to the amplifying process if the pump source was not sufficiently coherent.

The stimulated Brillouin scattering process is presently considered a deterrent to very coherent light transmission in a fiber because the incident light wave is easily converted to the Brillouin shifted light wave.

Stimulated Brillouin scattering or the Brillouin laser will, however, be important for optical processing based on very coherent light waves, such as generation of light waves with closely separated frequencies.

A great advantage of the acoustic wave, in comparison to the optical mode of lattice vibration, is that it can be generated and amplified by certain electronic processes in semiconductor crystals. For example, Nishizawa and Okamoto showed that an ultrasonic wave was generated following Gunn diode oscillations in GaAs [43]. Also, Nishizawa, Suzuki, and Suto carried out an experiment to generate a frequency-shifted light wave by impinging a microwave frequency sound onto a CdS crystal [44]. Although a practical semiconductor Brillouin laser is not yet available, an appropriate interaction process in a semiconductor will enable a low-threshold Brillouin laser to be realized in future.

1.3 COMPONENTS AND SYSTEMS IN WIDEBAND OPTICAL COMMUNICATIONS

The present optical communication systems are fundamentally based on three components: a coherent light source (i.e., a laser diode), an optical fiber as the light transmission medium, and a demodulator or detector of the light signal. The pin photodiode as well as the avalanche photodiode and Schottky diode are used for detection of high-frequency light signals. In present-day optical communication systems, the light pulse repetition rate is on the order of 1 Gb/s. The maximum rate for next-generation optical communication systems is expected to exceed 10 Gb/s. Even so, this is only 1 part in 10,000 of the light frequency, which is about 300 THz. That is to say, we are presently not fully utilizing the tremendously large information capacity of light.

Each of the three components has its own dynamic limitations. Although the pin photodiode has the highest response speed among all kinds of the efficient solid-state light detectors, its ultimate speed is limited to less than several tens of gigahertz because of carrier diffusion that occurs during drifting of carriers in the depletion layer of the pin junction. Even for a GaAs Schottky diode, a maximum frequency is approximately 100 GHz.

On the other hand, the laser diodes are directly modulated through the injection current. The maximum modulation speed is presently restricted by a relaxation effect. The carrier recombination lifetime changes with an increase in the optical field. As a result, the relaxation oscillations of the light intensity occur, and the modulation efficiency is steeply reduced at frequencies higher than the relaxation frequency, which is several tens of gigahertz in practice. For direct modulation, there is another problem called frequency chirping. The light frequency changes over several tens or even several hundred gigahertz during the duration of the pulse as a result of the change in the refractive index caused by injected carriers and by changes in the instantaneous temperature. These problems will be overcome by external modulation methods, although presently their efficiency is not sufficient. It was estimated that the ultimate limit of the modulation frequency can be almost 1 THz if the phase velocities of the modulation electric field and the light field are matched in a $LiNbO_3$ waveguide modulator [45]. As for the optical fiber, the optical pulse shape is broadened with increasing transmission distance as a result of the frequency dispersion of the light velocity caused by the dispersion of the refractive index and the waveguide structure. There are two different approaches for future optical communication that will increase the information capacity to be much greater than at present. One is based on frequency division, the other on time division. We call the first wideband optical communications and the latter high-speed optical communications.

Let us consider a terahertz-band optical communication in which the whole modulation bandwidth of the signal laser light extends over 1 THz , as illustrated in Figure 1.3.

Each channel will be pulse-modulated with a bandwidth of 1 to 10 GHz, which corresponds to the present level of the pulse modulation speed. However, between hundreds and several thousand channels with closely separated optical frequencies ω_1, ω_2, ... ω_i, ... ω_n with a frequency interval $\Delta\omega$ will then exist in the whole bandwidth. Or instead, we can assume light signals from a number of laser diodes each having a different optical frequency are transmitted through an optical fiber. The semiconductor Raman laser could act as a demodulator in such a wideband optical communication [46,47]. Each pulse with any one of these different optical frequencies can be discriminated by using a semiconductor Raman laser, provided the frequency interval is larger than the limit determined by the dispersion of the light velocity in the fiber transmission line. The semiconductor Raman laser is essentially a frequency-selective light amplifier. The amplified light frequency is determined by the pump light frequency, ω_L, and is given by

$$\omega_L = \omega_i + \omega_{ph}$$

If the pump source is a laser diode, we can easily tune the pump frequency, ω_L, to select an appropriate signal frequency, ω_i, by simple adjustment of the diode

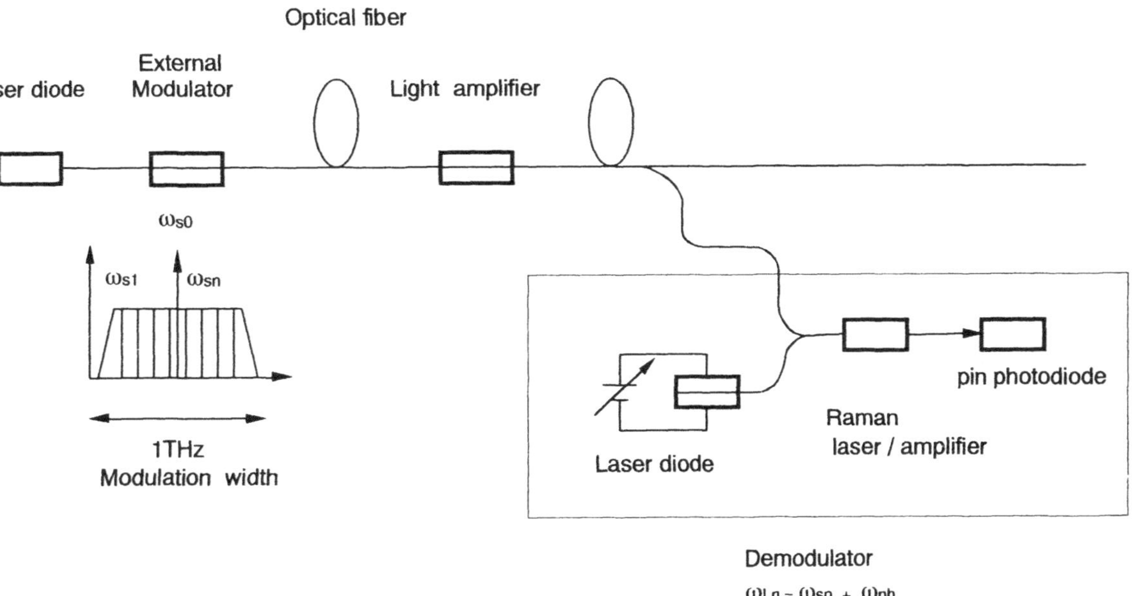

Figure 1.3 Heterodyne-type demodulation using the semiconductor Raman laser in a wideband optical communication system.

current. The selected component is finally detected by a pin photodiode. This is a heterodyne-type demodulation method in a sense that the frequency discrimination is made by a local light oscillator. It differs, however, from the conventional optical heterodyne method in which light mixing is carried out by a detector pin photodiode.

An important parameter of a heterodyne-type demodulator is the demodulator bandwidth $\Delta\omega$. If $\Delta\omega$ is made small, the number of demodulators necessary for the demodulation of the whole band will increase. Therefore, the bandwidth should be made as large as possible within the limitation arising from the photodiode and electronic amplifier bandwidths. The semiconductor Raman laser amplifier bandwidth is several tens of gigahertz. It is thought to be very suitable from this point of view.

As another requirement, frequency tuning must be made rapidly and in a simple manner. The semiconductor Raman laser, if it can be pumped by a laser diode, will enable rapid-tuning over several terahertz by simple adjustment of the laser diode current. Therefore, it is very important to develop a semiconductor Raman laser with a low-threshold pump power, much less than the 1 W level.

In comparison to frequency division, time division is limited to below several tens of gigahertz if we only consider a simple technological extension of present optical communications. Hasegawa and Tappert proposed soliton propagation in an optical fiber [48]. It was demonstrated that femtosecond optical pulses could be transmitted in a long-distance fiber. Fundamentally, however, a few problems remain to be solved. Generation of optical pulses with extremely high speed is necessary. To access the wideband capability, mode-locking of a laser diode cannot be applied because it is a resonance effect with a narrow band. There is presently no efficient light detector that has a short enough response time.

For long-distance optical communication systems, the light amplifiers at the intermediate stations (inline repeater amplifier) as well as the postamplifiers, which are used for the amplification of the output pulses from a signal source laser diode, are important. There are presently two kinds of amplifiers. One is the rare-earth doped fiber amplifier, in particular the erbium-doped fiber. The other is the laser diode amplifier. It is interesting to compare these amplifiers with the semiconductor Raman laser amplifier.

For an inline repeater amplifier, the bandwidth must be large enough to amplify all the frequency components or all the laser lines. Therefore, it is desirable to have a bandwidth greater than 1 THz. The rare-earth doped fiber amplifier and the laser diode amplifier have enough bandwidth. In contrast, the Raman laser amplifier can pick up precisely one component from a number of frequency components.

On the other hand, it should be noted that the amplifier noise increases proportionally with increase in the bandwidth. Therefore, the preamplifier that is used in a demodulator must have a narrow bandwidth that is coincident with the detector bandwidth and the electronic amplifier bandwidth. Therefore, the semiconductor Raman laser is very suitable as a light amplifier in a demodulator system.

REFERENCES

[1] Y. Watanabe and J. Nishizawa, "Semiconductor Maser," Japanese Patent 273217, Application 1957.

[2] J. Nishizawa, "Solid State Circuit with Optical Directivity," Japanese Patent 762975, Application 1960.

[3] J. Nishizawa, "History and Characteristics of Semiconductor Laser" (in Japanese), *Densi Kagaku*, Vol. 14, 1963, pp. 17–20.

[4] J. Nishizawa, "Esaki Diode and Long Wavelength Laser" (in Japanese), *Denshi Gijutsu*, Vol. 7, 1965, pp. 101–106.

[5] J. Nishizawa and Y. Watanabe, "Semiconductor Devices with High Resistive Regions," Japanese Patent 205068, Application 1950.

[6] J. Nishizawa and Y. Watanabe, "Semiconductor Optoelectronic Converters with High Resistive Thin Layers," Japanese Patent 221218, Application 1953.

[7] J. Nishizawa and Y. Watanabe, "Reverse Bias Characteristics of a Semiconductor Rectifiers," *Record of Electrical and Communication Engineering Conversation*, Tohoku University, Vol. 21, 1952, pp. 37–40.

[8] E.J. Woodburg and W.K. Ng, "Ruby Laser Operation in the Near IR," *IRE Proc.*, Vol. 50, 1962, p. 2367.

[9] F. Inaba, H. Itoh and N. Tanno, "Nonlinear Optics" (in Japanese), *Laser Handbook*, ed. by C. Yamanaka, Tokyo: OHMSHA Ltd, pp. 149–161.

[10] N. Bloembergen, G. Bret, P. Lallemand, A. Pine and P. Simova, "Controlled Stimulated Raman Amplification and Oscillation in Hydrogen Gas," *IEEE J. Quant. Electron.*, Vol. QE3, 1967, p. 197.

[11] O. Rahn, M. Maier and W. Kaiser, *Opt. Comm.*, Vol. 1, 1969, p. 109.

[12] G. Bret and M. Denariez, *J. Chem. Phys.*, Vol. 64, 1967, p. 222.

[13] J.B. Grun, A.K. McQuillan and B.P. Stoicheff, "Intensity and Gain Measurements on the Stimulated Raman Emission in Liquid O_2 and N_2," *Phys. Rev.*, Vol. 180, 1969, pp. 61–68.

[14] J.G. Skinner and W.G. Nilsen, "Absolute Raman Scattering Cross-Section Measurement of the 992 cm^{-1} Line of Benzene," *J. Opt. Soc. Amer.*, Vol. 58, 1968, pp. 113–119.

[15] F.J. McClung and D. Weiner, "Measurement of Raman Scattering Cross Sections for Use in Calculating Stimulated Raman Scattering Effects," *J. Opt. Soc. Amer.*, Vol. 54, 1964, pp. 641–643.

[16] F.J. McClung, W.G. Wagner and D. Weiner, "Mode Structure Independence of Stimulated Raman-Scattering Conversion Efficiencies," *Phys. Rev. Lett.*, Vol. 15, 1965, pp. 96–97.

[17] W.R.L. Clements and B.P. Stoicheff, "Raman Linewidths for Stimulated Threshold and Gain Calculations," *Appl. Phys. Lett.*, Vol. 12, 1968, pp. 246–248.

[18] A.K. McQuillan, W.R.L. Clements and B. P. Stoicheff, "Stimulated Raman Emission in Diamond: Spectrum, Gain, Angular Distribution of Intensity," *Phys. Rev.*, Vol. A1, 1970, pp. 628–635.

[19] G. Bisson and G. Mayer, "Effects Raman Stimules dans la Calcite," *J. de Phys.*, Vol. 29, 1968, pp. 97–110.

[20] W.D. Johnston, Jr., I.P. Kaminov and J.G. Bergman, Jr. "Stimulated Raman Gain Coefficients for Li_6NbO_3," *Appl. Phys.*, Lett. Vol. 13, 1968, pp. 190–193.

[21] J.M. Raslton and R.K. Chang, "Spontaneous Raman-Scattering Efficiency and Stimulated Scattering in Silicon," *Phys. Rev.*, Vol. B2, 1970, pp. 1858–1862.

[22] C.K.N. Patel, E.D. Shaw and R.J. Kerl, "Tunable Spin-Flip Laser and Infrared Spectroscopy," *Phys. Rev. Lett.*, Vol. 25, 1970, pp. 8–11.

[23] N. Bloembergen, *Nonlinear Optics*, New York: W. A. Benjamin, 1965.

[24] E. Garmire, F. Pandaresse and C. H. Townes, "Coherently Driven Molecular Vibrations and Light Modulation," *Phys. Rev. Lett.*, Vol. 11, 1963, pp. 160–163.
[25] N. Bloembergen and Y.R. Shem, "Quantum-Theoretical Comparison of Nonlinear Susceptibilities in Paramagnetic Media, Lasers and Raman Lasers," *Phys. Rev.*, Vol. 133, 1964, pp. A37–A49.
[26] J. Nishizawa and K. Suto, "Semiconductor Raman Laser," *J. Appl. Phys.*, Vol. 51, 1980, pp. 2429–2431.
[27] K. Suto, S. Ogasawara, T. Kimura and J. Nishizawa, "Buried-Heterostructure Semiconductor Raman Laser with Threshold Pump Power Less Than 1W," *J. Appl. Phys.*, Vol. 66, 1989, pp. 5151–5155.
[28] C.K.N. Patel and E.D. Shaw, "Tunable Stimulated Raman Scattering From Conduction Electrons in InSb," *Phys. Rev. Lett.*, Vol. 24, 1970, pp. 451–455.
[29] C.R. Pidgeon, B. Lax, R.L. Aggarwal, C.E. Chase and F. Brown, "Tunable Coherent Radiation Source in the 5-m Region," *Appl. Phys. Lett.*, Vol. 19, 1971, pp. 333–335.
[30] W.R. Johnston and I.P. Kaminow, "Temperature Dependence of Raman and Rayleigh Scattering in $LiNbO_3$ and $LiTaO_3$," *Phys. Rev.*, Vol. 168, 1968, pp. 1045–1054.
[31] J.M. Yarborough, S.S. Sussman, H.E. Pulhoff, R.H. Pantel and B.C. Johnson, "Efficient, Tunable Optical Emmission From $LiNbO_3$ Without a Resonator," *Appl. Phys. Lett.*, Vol. 15, 1969, pp. 102–105.
[32] R.H. Stolen, E.P. Ippen and A.R. Tynes, "Raman Oscillation in Glass Optical Waveguides," *Appl. Phys. Lett.*, Vol. 20, 1972, pp. 62–64.
[33] R.Y. Chiao, C.H. Townes and B.P. Stoicheff, "Stimulated Brillouin Scattering and Coherent Generation of Intense Hypersonic Waves," *Phys. Rev. Lett.*, Vol. 12, 1964, pp. 592–595.
[34] E.E. Hagenlocker, R.W. Minck and W.G. Rado, "Effects of Phonon Lifetime on Stimulated Optical Scattering in Gases," *Phys. Rev.*, Vol.154, 1966, pp. 226–233.
[35] R.Y. Chiao and P.A. Fleury, *Physics of Quantum Electronics*, ed. P.L. Kelly, B. Lax and P.E. Tannenwald, New York: McGraw-Hill, 1966, p. 241.
[36] A. Laubereau, W. Englisch and W. Kaiser, "Hypersonic Absorption of Liquids Determined from Spontaneous and Stimulated Brillouin Scattering," *IEEE J. Quant. Electron.*, Vol. QE5, 1969, p. 410–415.
[37] D. Pohl and W. Kaiser, "Time-Resolved Investigation of Stimulated Brillouin Scattering in Transparent and Absorbing Media: Determination of Phonon Lifetimes," *Phys. Rev.*, Vol. B4, 1970, pp. 31–43.
[38] V.S. Starunov and I.L. Fabelinskii, *Uspeki fiz. Nauk*, Vol. 98, 1969, p. 441.
[39] M. Denariez and G. Bret, "Investigation of Rayleigh Wings and Brillouin-Stimulated Scattering in Liquids," *Phys. Rev.*, Vol. 171, 1968, p. 160.
[40] J. Walder and C.L. Tang, "Photoelastic Amplification of Light and the Generation of Hypersound by the Stimulated Brillouin Process," *Phys. Rev. Lett.*, Vol 19, 1967, pp. 623–626.
[41] E.P. Ippen and R.H. Stolen, "Stimulated Brillouin Scattering in Optical Fibers," *Appl. Phys. Lett.*, Vol. 21, 1972, pp. 539–541.
[42] N.A. Olssen and J.P. Van Der Ziel, "Characteristics of a Semiconductor Laser Pumped Brillouin Amplifier with Electronically Controlled Bandwidth," *J. Lightwave Technol.*, Vol. LT-5, 1987, pp. 147–153.
[43] K. Okamoto, H. Ishikawa and J. Nishizawa, "Ultrasonic Wave Generation in Oscillating n-GaAs," *RIEC Technical Report*, Tohoku University: Research Institute of Electrical Communication, Vol. TR-22, 1967, pp. 1–8.
[44] S. Suzuki, J. Nishizawa and K. Suto, "The Coherent Interaction of Externally Generated 35-GHz Sound With the Light in CdS," *Appl. Phys. Lett.*, Vol. 30, 1977, pp. 310–312.
[45] M. Minakata and T. Chattopadhyay, "High Frequency $LiNbO_3$ Optical Modulators," *ERATO Summary Reports of NISHIZAWA Terahertz Project*, 1992, pp. 47–58.

[46] J. Nishizawa and K. Suto, "Lightwave Demodulator," Japanese Patent 1605283, Application 1981.
[47] K. Suto and J. Nishizawa, "Characteristics of the Epitaxial Semiconductor Raman Laser," *IEE Proc.*, Vol. 133, Pt. J, 1986, pp. 259–263.
[48] A. Hasegawa and F. Tappert, "Transmission of Stationary Nonlinear Optical Pulses in Dispersive Dielectric Fibers .1 Anomalous Dispersion," *Appl. Phys. Lett.*, Vol. 23, 1973, pp. 142–144.

Chapter 2
Principles of Raman Laser Operation

This chapter first discusses lattice vibrations with optical modes, which are known to cause the lattice reflection band in an infrared wavelength region. Then, it addresses Raman scattering in terms of the nonlinear polarization induced by the lattice vibrations. It should be noted that the Raman polarizability in III-V compound semiconductors is very large compared to those in other kinds of materials, such as II-VI compounds, ionic crystals, or SiO_2. The Stokes light is observed at a frequency $\omega_s = \omega_L - \omega_{ph}$, where ω_L is the frequency of the pump laser light, and ω_{ph} is the frequency of the lattice vibration. When the intensity of the pump light is weak, only spontaneous Raman scattering is observed. However, when the pump light intensity is strong enough, stimulated Raman scattering becomes dominant, and thus a coherent light wave introduced into the material can be amplified. The light frequency to be amplified is not fixed, as in other kinds of light amplifiers, but is dependent on the frequency of the pump light. Frequency selectivity is the unique characteristic of the Raman laser/amplifier. The light amplification can be analyzed as electromagnetic wave propagation based on Maxwell equations in a medium with nonlinear polarization.

If the crystal is placed between a pair of high-reflectivity mirrors constructing a resonator, coherent light oscillation is realized. We discuss the optical field in the resonator. In contrast to the laser diode, there is no strong optical absorption at the amplifying frequency. In ideal circumstances, where there is no internal optical absorption, complete power conversion from the pump light to the Stokes light is expected. Finally, we discuss Raman amplification in optical fibers. Although the Raman polarizability of silica fiber is very small, significant light amplification can be obtained after long interaction distances with the pump light.

2.1 LATTICE VIBRATION IN SEMICONDUCTORS

We consider the lattice vibration in a compound semiconductor in a linear oscillator scheme. Let $\mathbf{Q} = \mathbf{u}^+ - \mathbf{u}^-$ be the lattice displacement with a unit cell of a polar crystal with a cubic symmetry, where $\mathbf{u}^+$ and $\mathbf{u}^-$ are the displacement vectors of the cation and anion constituting the sublattices of the crystal. If we assume the presence of an electric field, **E**, acting on the ions, the equation of motion can be described as follows:

$$\frac{d^2\mathbf{Q}}{dt^2} = -\omega_0^2\mathbf{Q} - \Gamma\frac{d\mathbf{Q}}{dt} + \frac{e_B^*}{M}\mathbf{E} \tag{2.1}$$

where ω_0 is the mechanical resonance frequency, which is equal to the transverse optical phonon frequency ω_{TO}, Γ is the lattice damping constant, and M is the reduced mass of the cation and anion in a unit cell. The effective charge e_B^* was introduced by Born and Huang [1] and can be expressed by using the dielectric constant at optical frequency ε_∞, and the static dielectric constant ε_s:

$$e_B^* = \left[\frac{M}{N}(\varepsilon_s - \varepsilon_\infty)\right]^{1/2}\omega_0 \tag{2.2}$$

where N is the number of unit cells in a unit volume.

We assume that the lattice displacement and the electric field vibrate with frequency ω and propagate with momentum **k**, written in the form:

$$\mathbf{Q} = \mathbf{Q}(\omega)\exp(j\mathbf{k}\cdot\mathbf{r} - j\omega t), \quad \mathbf{E} = \mathbf{E}(\omega)\exp(j\mathbf{k}\cdot\mathbf{r} - j\omega t). \tag{2.3}$$

Then, the equation of motion becomes:

$$(\omega_0^2 - \omega^2 - j\omega\Gamma)\mathbf{Q}(\omega) = \frac{e_B^*}{M}\mathbf{E}(\omega) \tag{2.4}$$

First, we consider the case when the electric field is that of an electromagnetic wave, described by the Maxwell equation:

$$\nabla^2\mathbf{E} - \frac{1}{c^2}\frac{\partial^2\mathbf{E}}{\partial t^2} = \frac{1}{c^2\varepsilon_0}\frac{\partial^2\mathbf{P}}{\partial t^2} \tag{2.5}$$

with

$$\mathbf{P} = Ne_B^*\,\mathbf{Q} + (\varepsilon_\infty - \varepsilon_0)\mathbf{E} \tag{2.6}$$

where **P** is the dielectric polarization. Using the expressions (2.3), the Maxwell equation becomes:

$$\mathbf{E} = -\frac{[\mathbf{k}(\mathbf{k}\cdot\mathbf{P}) - \omega^2\mathbf{P}/c^2]}{\varepsilon_0(k^2 - \omega^2/c^2)} \tag{2.7}$$

For a transverse plane wave, we put $(\mathbf{k}\cdot\mathbf{P}) = 0$. Then the combination of (2.7) with (2.2) and (2.4) gives the following relation:

$$\begin{aligned}\varepsilon_0\frac{c^2k^2}{\omega^2} \equiv \varepsilon = \varepsilon_\infty + (\varepsilon_s - \varepsilon_\infty)\frac{\omega_0^2}{\omega_0^2 - \omega^2 - j\omega\Gamma} \doteqdot \varepsilon_\infty + (\varepsilon_s - \varepsilon_\infty)\frac{\omega_0^2}{\omega_0^2 - \omega^2} \\ = \frac{\varepsilon_s\omega_0^2 - \varepsilon_\infty\omega^2}{\omega_0^2 - \omega^2}\end{aligned} \tag{2.8}$$

This equation is the dispersion relation for the coupled mode of lattice vibration and electromagnetic wave, which is illustrated in Figure 2.1. There are two branches in the dispersion relation, and the lower frequency branch is called the polariton mode.

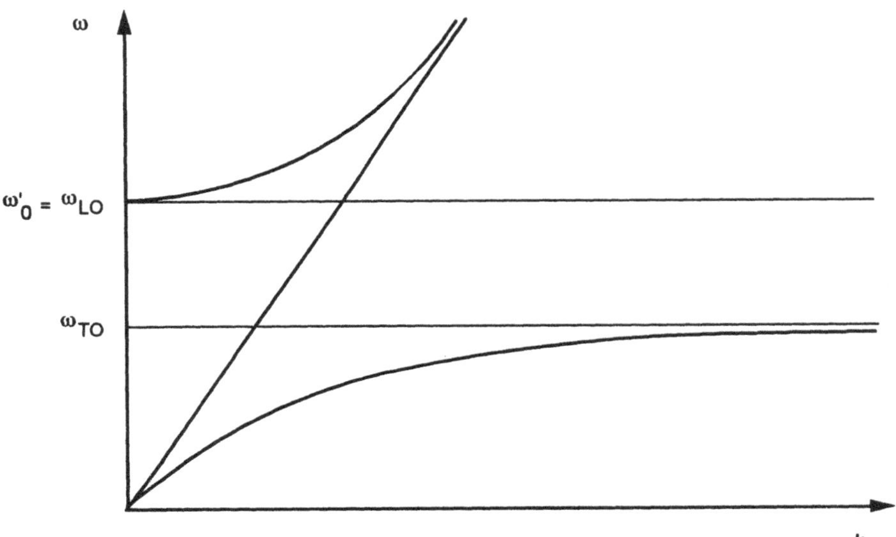

Figure 2.1 Dispersion curve of long-wavelength optical mode vibrations.

On the other hand, the upper branch gives a finite frequency, ω_0', in the limit of $\mathbf{k} = 0$, given by:

$$\omega_0' = \sqrt{\frac{\varepsilon_s}{\varepsilon_\infty}}\,\omega_0 \tag{2.9}$$

This frequency is the same as the longitudinal optical mode frequency, ω_{LO}, which will be further discussed.

For the longitudinal optical mode lattice vibration, the electric field induced by the dielectric polarization is given by the condition that there is no net charge:

$$\mathbf{D} = \varepsilon_0\mathbf{E} + \mathbf{P} = 0 \tag{2.10}$$

Then (2.4) takes the following form:

$$\left(\frac{\varepsilon_s}{\varepsilon_\infty}\,\omega_0^2 - \omega^2 - j\omega\Gamma\right)\mathbf{Q}(\omega) = 0 \tag{2.11}$$

Therefore, it is seen that the longitudinal optical mode vibration occurs at a frequency $\omega_{LO} = \sqrt{\varepsilon_s/\varepsilon_\infty}\,\omega_0$, which is the same as (2.9). However, this time the resonance frequency is independent of the wave vector $\mathbf{k}$, as illustrated in Figure 2.1, provided $\mathbf{k}$ is so small that we can assume the lattice is a continuum.

It should be noted that the electric field, $\mathbf{E}$, adopted here is the macroscopic field but not the local electric field, $\mathbf{E}_{loc}$, acting on individual ions, which is considered to be given by:

$$\mathbf{E}_{loc} = \mathbf{E} + \frac{1}{3\varepsilon_0}\mathbf{P} \tag{2.12}$$

The lattice resonance is observed as a lattice reflection band or "Reststrahlen" band. The reflectance, R, for far-infrared light incident normally on the surface of a crystalis is given by:

$$R = \left|\frac{\sqrt{\varepsilon/\varepsilon_0} - 1}{\sqrt{\varepsilon/\varepsilon_0} + 1}\right|^2 \tag{2.13}$$

where ε is given by (2.8), which is generally a complex number. If we neglect the lattice damping term, (2.8) shows that ε has negative value in the frequency range

$\omega_0 < \omega < \omega_0'$. This means that the reflectance, R, is equal to 1 in this frequency range. In reality, R is less than 1 as a result of lattice damping.

Hass [2] and others made a detailed analysis of the lattice reflection band in III-V compound semiconductors. Figure 2.2 shows the lattice reflection band of GaP measured by Kleinmanand Spitzer [3]. The solid line is the calculation with the assumption $\Gamma/\omega_0 = 0.003$.

Table 2.1 gives the transverse and longitudinal optical phonon frequencies, $v_{TO} = \omega_{TO}/2\pi$ (THz), $v_{LO} = \omega_{LO}/2\pi$ (THz), and the ratio of the damping constant, $\gamma = \Gamma/2\pi$ (GHz) to v_{TO}, for various III–V compound semiconductors [2]. The lattice damping constant for GaP is the smallest among those for the III–V compounds.

As is given in Table 2.2, the lattice damping constant for GaP is the smallest when compared to those of the II–VI compounds, ionic crystals, and the relevant damping parameters for liquids measured by Raman scattering. This is one of the reasons why low-threshold semiconductor Raman lasers can be realized using GaP.

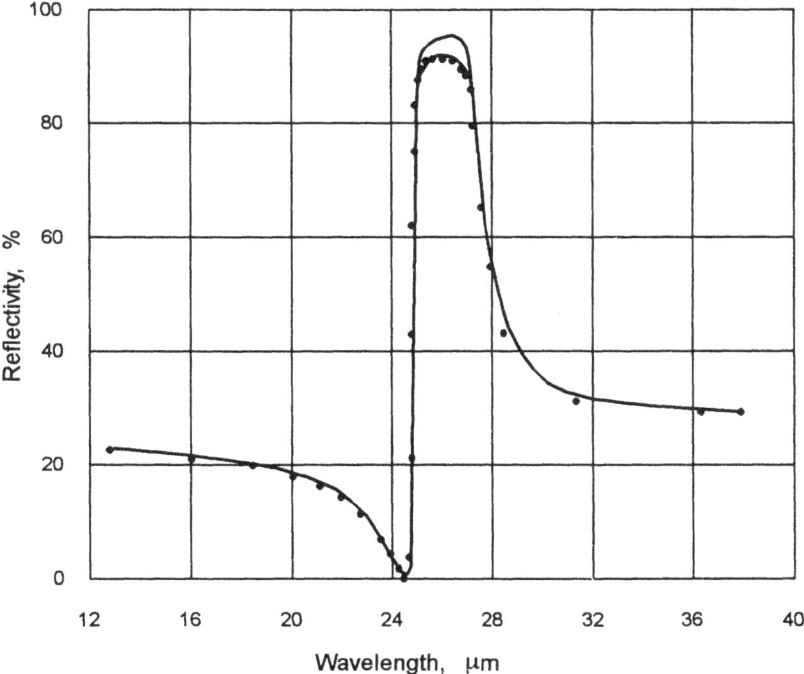

Figure 2.2 Lattice reflection spectrum of GaP (data shown by points and the calculated fit by the solid curve) [3].

Table 2.1
Optical Phonon Frequencies in III–V Compounds [2]

Compound	*Type of Measurement*	*Temperature*	ν_{TO} *(THz)*	ν_{LO} *(THz)*	γ/ν_{TO}
BN	R	Room	31.9	40.2	
BP	R	Room	24.6	25.0	
AlP	T	300 K	13.2	15.0	
AlAs	T	300 K	10.2	12.1	
AlSb	R	300 K	9.557	10.18	0.0059
GaP	Raman	Room	11.01	12.08	0.003
	R	300 K	10.98	12.05	
	Raman	Helium	10.96	12.08	
GaAs	R	295 K	8.04	8.739	0.007
	R	Helium	8.193	8.913	
GaSb	R	Helium	6.910	7.204	<0.01
InP	R	Room	9.210	10.45	0.01
	Raman	Room	9.105	10.34	
	Raman	Helium	9.24	10.48	
InAs	R	Helium	6.562	7.294	<0.01
InSb	R	300 K	5.402	5.73	<0.015

R = reflection, T = transmission.

Table 2.2
Damping Constants of Optical Mode Lattice Vibrations

Material	$\Gamma/2\pi$ *(GHz)*	*Reference*
GaP	36	[2]
GaAs	60	[2]
InP	90	[2]
CdS	189 (TO_{11})	[4]
	225 (LO_{11})	
$LiNbO_3$	600	[5]
$LiTaO_3$	660	[5]
C_6H_6	69	[6]
CS_2	42	[7]

2.2 RAMAN SCATTERING

Raman scattering is caused by a nonlinear interaction between the light field and the lattice vibration. Figure 2.3 shows two typical configurations for the measurements of spontaneous Raman scattering in semiconductors and other materials. The pump source is usually a high-intensity laser, and its wavelength is in the infrared-to-visible wavelength region for which the material is transparent. In Figure 2.3(a), Raman scattering is observed in a direction perpendicular to the pump beam direction. Usually, spontaneous Raman scattering is measured in this configuration to avoid the strong pump light coming into the spectrometer. However, the parallel configuration shown in Figure 2.3(b) is also used to measure a lattice polariton scattering or a collinear interaction between light and lattice vibrations, as in Raman amplification.

Spontaneous Raman scattering occurs with the generation or annihilation of phonons, each process resulting in Stokes or anti-Stokes radiation, respectively. If we let the frequencies of the pump beam, Stokes beam, anti-Stokes beam, and phonon be ω_L, ω_s, ω_a, and ω_{ph}, respectively, they should satisfy the relations:

$$\omega_L = \omega_s + \omega_{ph} \tag{2.14}$$

$$\omega_L = \omega_a - \omega_{ph} \tag{2.15}$$

The origin of the Raman scattering can be ascribed to the nonlinear term of the polarizability α_{ij}. The polarization **P** of a material is given by:

$$\mathbf{P}_i = \alpha_{ij}\mathbf{E}_j \tag{2.16}$$

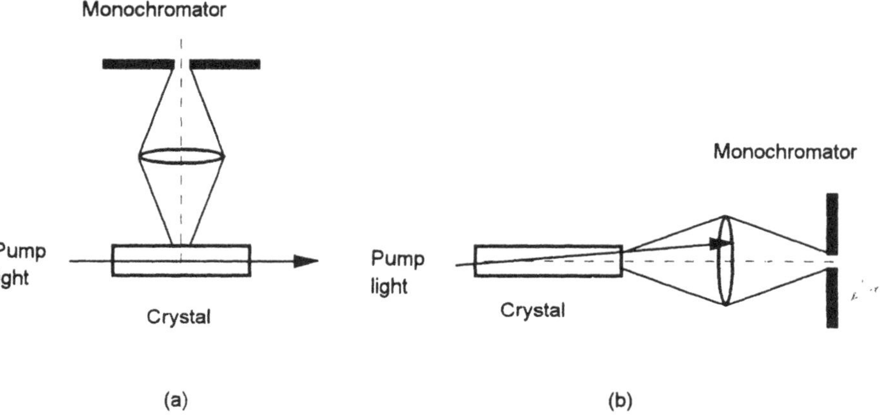

Figure 2.3 Arrangements for Raman scattering measurements: (a) perpendicular and (b) parallel.

where suffixes i and j refer to the directions of components of polarization and electric field vectors, respectively. Following the Einstein summation convention, when a suffix occurs twice in the same time, summation with respect to this suffix is to be made; that is, (2.16) represents $\mathbf{P}_i = \Sigma_j \alpha_{ij}\mathbf{E}_j$. Here we have defined α_{ij} for a unit volume but not for a unit cell, so that α_{ij} is the same as the susceptibility x_{ij}, which is usually used when the macroscopic properties are to be described. The Raman polalizability, α_{ijk}, is defined as the change in α_{ij} via the change in the relative displacement of the cation and anion in a unit cell, $\mathbf{Q} = \mathbf{u}^+ - \mathbf{u}^-$:

$$\alpha_{ijk} = \frac{\partial \alpha_{ij}}{\partial Q_k} \tag{2.17}$$

where Q_k is a component of $\mathbf{Q}$ in direction k. The nonlinear part of the polarization, $\mathbf{P}^{NL}$, is therefore given by:

$$\mathbf{P}_i^{\mathbf{NL}} = \alpha_{ijk}\mathbf{E}_j Q_k \tag{2.18}$$

The intensity of the spontaneous Raman scattering is usually described by the scattering efficiency $S/\ell d\Omega$, where S is the fraction of the scattered power relative to the pump power into a solid angle $d\Omega$ for an optical path length ℓ. If the pump light is unpolarized and the scattered light is not analyzed, the scattering efficiency for the Stokes radiation is given by [1,8]:

$$S = \frac{3\hbar\omega_s^4 l d\Omega}{\rho c^4 \omega_{ph}} \frac{|\alpha_{ijk}|^2}{16\pi^2\varepsilon_0^2}(n_0 + 1) \tag{2.19}$$

where ρ is the density of the crystal, and n_0 is the Bose factor for phonons given by:

$$n_0 = \frac{1}{\exp\left(\dfrac{\hbar\omega_{ph}}{kT}\right) - 1} \tag{2.20}$$

If we replace $(n_0 + 1)$ in (2.19) by n_0, we obtain the expression of the scattering efficiency for the anti-Stokes radiation.

In general, however, we must take into account the fact that both the pump and scattered radiations are polarized. Let their polarization vectors be $\mathbf{e}_L$ and $\mathbf{e}_s$, respectively. Also, we define a unit vector $\boldsymbol{\xi}$, which is in the direction of the displacement of the phonon. Then, the scattering efficiency can be expressed in the form:

$$S = A\left[\sum_{ijk=x,y,z} e_L^i\, \alpha_{ijk}(\omega_{ph})\boldsymbol{\xi}^k . e_s^j\right]^2 \tag{2.21}$$

The Raman polarizability tensor α_{ijk} depends on ω_L, ω_s (or ω_A), and ω_{ph}, but we have only explicitly included ω_{ph} because α_{ijk} has a sharp frequency dependence around the lattice resonance frequencies. Cubic compound semiconductors like III-V compounds and II-VI compounds belong to the T_d crystal symmetry group. Because they are lacking in inversion symmetry, there are three nonzero components with the same value in the third-rank tensor α_{ijk}:

$$\alpha_{xyz} = \alpha_{zxy} = \alpha_{yzx} \tag{2.22}$$

Let us consider the measurement using the collinear configuration shown in Figure 2.3(*b*). Figure 2.4 shows spontaneous Raman scattering spectra in GaP for the two typical crystal orientations [9]. In Figure 2.4(*a*), the pump and scattered radiations travel in the $\langle 001\rangle$ crystal orientation. From (2.20), we can understand that only the longitudinal optical phonon with $\boldsymbol{\xi}//\langle 001\rangle$ is excited, and, if we choose $\mathbf{e}_L//\langle 001\rangle$, we find that $\mathbf{e}_s//\langle 010\rangle$. Actually, a weak contribution from the transverse optical phonon is also observed at the side of the strong longitudinal phonon line. This may be due to the fact that we collect the radiation from a finite cone with a solid angle $\Delta\Omega$ around the light path.

If we set $\mathbf{e}_L//\langle 110\rangle$ instead of $\mathbf{e}_L//\langle 001\rangle$ while retaining $\boldsymbol{\xi}//\langle 001\rangle$, we find that $\mathbf{e}_s//\langle 110\rangle$, (i.e., parallel to $\mathbf{e}_L$).

In Figure 2.4(*b*), the pump and scattered radiations travel in the $\langle 110\rangle$ direction. Only the transverse optical phonons are excited; that is, $\boldsymbol{\xi}//\langle 1\bar{1}0\rangle$. If we choose $\mathbf{e}_L//\langle 001\rangle$, we have $\mathbf{e}_s//\langle 1\bar{1}0\rangle$. The observed transverse optical phonon band is composed of the polariton mode TO_f, which appears for the forward scattering, and also composed of the pure transverse optical phonon mode TO_b, which appears for the backward scattering. The latter appears as a result of multiple reflections of the pump and scattered radiations at the crystal surfaces.

The observed spectral broadening of the polariton mode is not due to the lattice damping but to the finite solid angle for detection $\Delta\Omega$. On the other hand, the spectral linewidth of the longitudinal optical phonon mode is not affected by $\Delta\Omega$ because there is little dispersion, as mentioned in Section 2.1. The linewidth for the *LO* phonon mode of GaP is roughly $\Delta\lambda \simeq 0.5$ Å at room temperature when the pump source is a 6,328 Å He-Ne laser. This value is in agreement with the results of lattice reflection measurements that gave the lattice damping factor $\Gamma \simeq 0.003\ \omega_0$ (i.e., $\Gamma/2\pi \simeq 30$ GHz).

In the case of crystals with T_d symmetry, the phonon displacement vector $\boldsymbol{\xi}$ has threefold degeneracy; that is, $\boldsymbol{\xi}$ can be any one of the $\langle 100\rangle$, $\langle 010\rangle$, or $\langle 001\rangle$ crystal axis or arbitrary linear combinations of them. To discuss more general cases

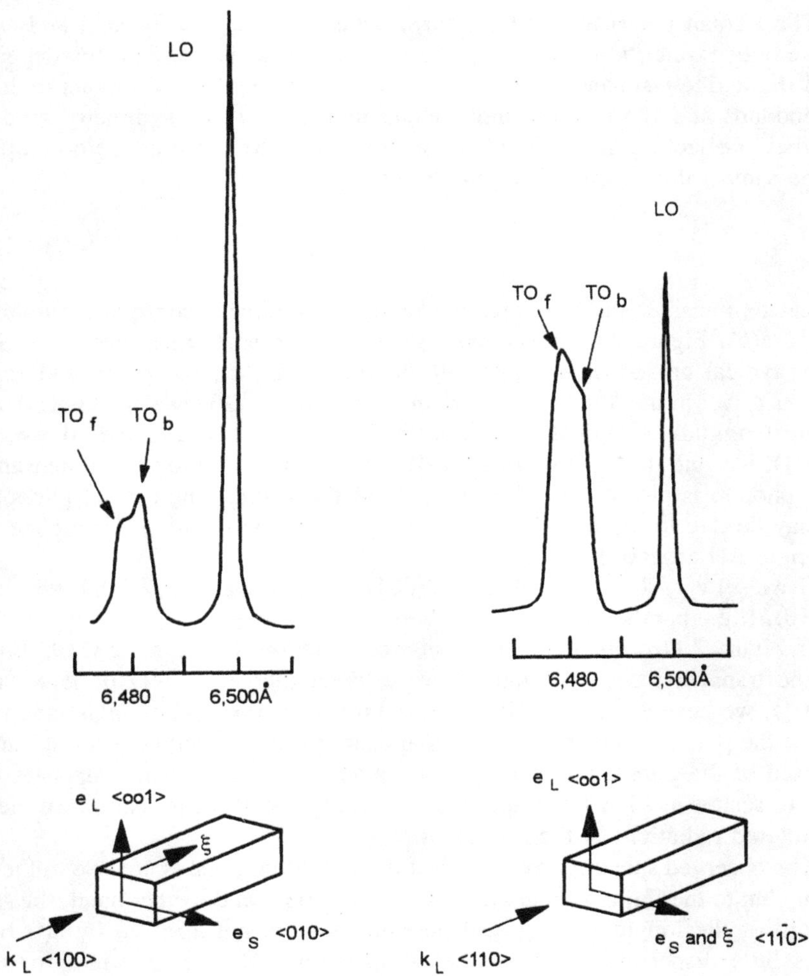

Figure 2.4 Spontaneous Raman scattering spectra in GaP with 6,328 Å incident radiation measured with the parallel arrangement [9].

with lower crystal symmetry, the scattering efficiency is often expressed in the following form, instead of (2.21):

$$S = A[\mathbf{e}_s R(m) \mathbf{e}_L]^2 \tag{2.23}$$

where $R(m)$ is called the Raman tensor for the mth igen mode of vibration. For the T_d symmetry group, $R(m)$ is expressed as follows:

$$R(x) = \begin{pmatrix} 0 & 0 & 0 \\ 0 & 0 & a \\ 0 & a & 0 \end{pmatrix}, R(y) = \begin{pmatrix} 0 & 0 & a \\ 0 & 0 & 0 \\ a & 0 & 0 \end{pmatrix}, R(z) = \begin{pmatrix} 0 & a & 0 \\ a & 0 & 0 \\ 0 & 0 & 0 \end{pmatrix} \tag{2.24}$$

The Raman tensor forms for other crystal symmetry groups are tabulated in [8].

The Raman polarizability in GaAs was measured by Cardona as a function of the pump wavelength as shown in Figure 2.5 [10].

From the value of a in Figure 2.5, the magnitude of the Raman polarizability for a unit volume, α_{ijk}, can be obtained by using the relation:

$$\alpha_{ijk} = \varepsilon_0 N a$$

The Raman polarizability is somewhat larger when the photon energy is close to the direct bandgap energy, but the increase is not as prominent as that related to higher energy transitions. Table 2.3 compares the Raman polarizabilities in various III–V compounds. These are the values in the nondispersion region below the bandgap.

For II–VI compounds, systematic data are not available. However, the Raman scattering efficiencies for LO phonons in CdS, ZnS, and ZnSe were measured to be only $1.7 \times 10^{-3} \sim 4 \times 10^{-2}$ times that of GaP at room temperature [12,13]. Other materials such as diamond, quartz, or ionic crystals also show smaller scattering efficiencies compared to those of the III–V compounds. Raman scattering efficiencies for ferroelectric materials, such as $LiTaO_3$ and $LiNbO_3$, are of the same order of magnitude as those of the III–V compounds [4].

However, it should be noted that the Raman gain is inversely proportional to the spontaneous Raman scattering linewidth Γ, which is equal to the lattice damping factor, as will be discussed later. As is given in Table 2.2, the Γ value for GaP is the smallest among those of the III–V compounds and is comparable to those of liquids such as CS_2, while the Γ values for $LiTaO_3$ and $LiNbO_3$ are almost an order of magnitude higher than that of GaP at room temperature.

Another important parameter for the Raman laser is the internal absorption loss. The absorption loss in GaP is extremely small, as will be discussed in Section 3.2. This should be related to the fact that GaP is an indirect bandgap material. All these factors make GaP the most suitable material for Raman laser oscillation and amplification.

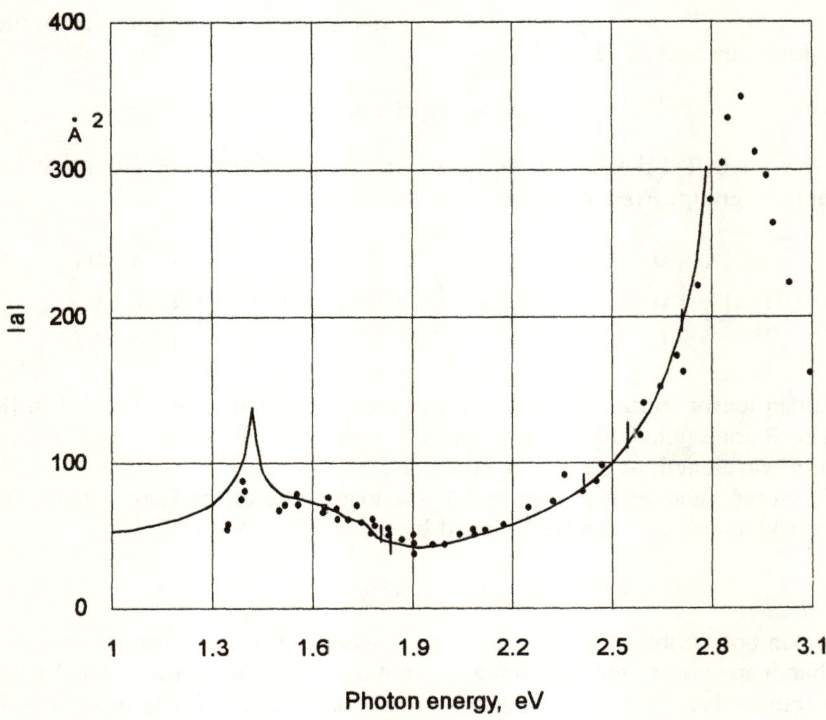

Figure 2.5 Raman tensor component a of GaAs as a function of photon energy at room temperature (data shown by points; the length of the vertical bars indicates estimated error, and the solid line is a theoretical fit) [10].

Table 2.3
Raman Polarizabilities of III–V Compounds in the Nonresonant Region Below the Energy Gap

Compound	*Raman Polarizability (A^2)*	*Reference*
AlSb	77	[11]
GaP	35	[11]
GaAs	55	[10]
GaSb	169	[11]
InP	41	[11]
InAs	84	[11]
InSb	206	[11]

2.3 LIGHT AMPLIFICATION BY STIMULATED RAMAN SCATTERING

2.3.1 Basic Theory

Light amplification by stimulated Raman scattering was first discussed by Bloembergen and Shen in terms of nonlinear susceptibilities [14,15]. We assume that the coherent optical fields of the pump light, $\mathbf{E}(\omega_L)$, and Stokes light, $\mathbf{E}(\omega_s)$, are present. We include the nonlinear polarization term into Maxwell equation expressed in the form:

$$\nabla^2\mathbf{E} - \frac{\varepsilon^\infty}{\varepsilon_0 c^2}\frac{\partial^2\mathbf{E}}{\partial t^2} = \frac{1}{\varepsilon_0 c^2}\frac{\partial \mathbf{P}^{NL}}{\partial t} \tag{2.25}$$

From (2.18), we have the following two propagation equations for $\mathbf{E}(\omega_L)$ and $\mathbf{E}(\omega_s)$:

$$\left(\nabla^2 + \frac{\omega_L^2}{c^2} n_L^2\right)\mathbf{E}_i(\omega_L) = -\frac{1}{\varepsilon_0 c^2}\omega_L^2\, \mathbf{P}_i^{NL}(\omega_L) = -\frac{1}{\varepsilon_0 c^2}\omega_L^2\, \alpha_{ijk}\mathbf{E}_j(\omega_s)\mathbf{Q}_k(\omega) \tag{2.26}$$

$$\left(\nabla^2 + \frac{\omega_s^2}{c^2} n_s^2\right)\mathbf{E}_i(\omega_s) = -\frac{1}{\varepsilon_0 c^2}\omega_s^2\, \mathbf{P}_i^{NL}(\omega_s) = -\frac{1}{\varepsilon_0 c^2}\omega_s^2\, \alpha_{ijk}\mathbf{E}_j(\omega_L)\mathbf{Q}_k^*(\omega) \tag{2.27}$$

where $\omega = \omega_L - \omega_s$. We have used the relation $\varepsilon = n^2$ between the dielectric constant ε and the refractive index n.

On the other hand, the nonlinear force term acting on the lattice points appears through the nonlinear polarization. This term can be derived by considering the nonlinear free energy density function F^{NL}, which is given by:

$$\mathrm{F}^{\mathrm{NL}} = -(\mathbf{P}^{NL}\cdot\mathbf{E}^* + c.c.) = -[\alpha_{ijk}\mathbf{E}_i(\omega_L)\mathbf{E}_j^*(\omega_s)\mathbf{Q}_k^*(\omega) + c.c.] \tag{2.28}$$

where $c.c.$ means the complex conjugate terms. Then, the nonlinear force $\mathbf{f}^{NL}$ is given by:

$$\mathbf{f}_k^{\mathrm{NL}}(\omega) = -\frac{1}{\mathrm{N}}\frac{\partial \mathrm{F}^{\mathrm{NL}}}{\partial \mathbf{Q}_k^*} = \frac{1}{\mathrm{N}}\alpha_{ijk}\mathbf{E}_i(\omega_L)\mathbf{E}_j^*(\omega_s) \tag{2.29}$$

Therefore, instead of (2.4), the equation of motion for the lattice displacement $\mathbf{Q}(\omega)$ becomes:

$$\mathrm{M}(\omega_0^2 - \omega^2 - j\omega\Gamma)\mathbf{Q}_k(\omega) = \frac{\mathrm{e}_{\mathrm{B}}^*}{\mathrm{M}}\mathbf{E}_k(\omega) + \frac{1}{\mathrm{N}}\alpha_{ijk}\mathbf{E}_i(\omega_L)\mathbf{E}_j^*(\omega_s) \tag{2.30}$$

For the longitudinal optical mode, (2.30) takes the following form, referring to (2.11):

$$\mathrm{M}(\omega_{LO}^2 - \omega^2 - j\omega\Gamma)\mathbf{Q}_k(\omega) = \frac{1}{\mathrm{N}}\,\alpha_{ijk}\mathbf{E}_i(\omega_L)\mathbf{E}_j^*(\omega_s) \tag{2.31}$$

Let us consider the same configuration as illustrated in Figure 2.3(*b*). The plane waves of the pump and the Stokes light travel along ⟨001⟩, with polarization directions $\mathbf{e}_L$//⟨100⟩ and $\mathbf{e}_s$//⟨010⟩, respectively, and the longitudinal optical vibrational mode with $\boldsymbol{\xi}$//⟨001⟩ is excited. Then, the propagation equations and the equation of motion take the following forms:

$$\left(\frac{\partial^2}{\partial z^2} + \frac{\omega_L^2}{c^2}\,n_L^2\right)E(\omega_L) = -\frac{1}{\varepsilon_0 c^2}\,\omega_L^2\alpha_{xyz}E(\omega_s)Q(\omega) \tag{2.32}$$

$$\left(\frac{\partial^2}{\partial z^2} + \frac{\omega_s^2}{c^2}\,n_s^2\right)E(\omega_s) = -\frac{1}{\varepsilon_0 c^2}\,\omega_s^2\alpha_{xyz}E(\omega_L)Q^*(\omega) \tag{2.33}$$

$$M(\omega_{LO}^2 - \omega^2 - j\omega\Gamma)Q(\omega) = \frac{\alpha_{xyz}}{N}\,E(\omega_L)E^*(\omega_s) \tag{2.34}$$

with $\omega = \omega_L - \omega_s$.

From these equations, we have the following equation for the Stokes field $E(\omega_s)$:

$$\begin{aligned}\left(\frac{\partial^2}{\partial z^2} + \frac{\omega_s^2}{c^2}\,n_s^2\right)E(\omega_s) &= -\frac{1}{\varepsilon_0 c^2}\,\omega_s^2\,\frac{\alpha_{xyz}^2}{MN(\omega_{LO}^2 - \omega^2 + j\omega\Gamma)}\,|E(\omega_L)|^2E(\omega_s)\\ &= -\frac{1}{\varepsilon_0 c^2}\,\omega_s^2\chi_s(\omega)|E(\omega_L)|^2E(\omega_s)\end{aligned} \tag{2.35}$$

where we have defined the nonlinear susceptibility $\chi_s(\omega)$ for the Stokes wave when the longitudinal mode is excited.

$$\chi_s(\omega) = \frac{\alpha_{xyz}^2}{MN(\omega_{LO}^2 - \omega^2 + j\omega\Gamma)} \tag{2.36}$$

The nonlinear polarization for the Stokes field $\mathbf{P}_s$ is given by:

$$\boldsymbol{P}_s = \chi_s|\boldsymbol{E}_L|^2\boldsymbol{E}_s$$

We first assume that the pump intensity is much higher than the Stokes intensity so that $E(\omega_L)$ is constant along the z direction; that is, we have two coupled equations (2.33) and (2.34). The solutions to these equations can be obtained by assuming the following simple forms:

$$E(\omega_L) = \frac{E_L}{\sqrt{2}} \exp(jk_L z - j\omega_L t) + c.c.$$

$$E(\omega_s) = \frac{E_s}{\sqrt{2}} \exp(jk_s z + \gamma_z - j\omega_s t) + c.c. \tag{2.37}$$

$$Q(\omega) = \frac{Q}{\sqrt{2}} \exp(jkz + \gamma_z - j\omega t) + c.c.$$

with $k = k_L - k_s$.

As a result, we have the following equation:

$$\left\{(jk_s + \gamma)^2 + \frac{\omega_s^2}{c^2} n_s^2\right\} E(\omega_s) = -\frac{\omega_s^2 \alpha_{xyz}^2}{\varepsilon_0 c^2 MN} \frac{|E(\omega_L)|^2}{\omega_{LO}^2 - \omega^2 + j\omega\Gamma} E(\omega_s) \tag{2.38}$$

where k_s satisfies $k_s = \omega_s / c\ n_s$.

We can neglect the γ^2 term in (2.38) because the electric field amplification factor γ should not be large. At the frequency of lattice vibration $\omega = \omega_{LO}$, γ is given by:

$$\gamma = \frac{\alpha_{xyz}^2}{2\varepsilon_0 cMN} \frac{1}{\omega_{LO}\Gamma} \left\{\frac{\omega_s}{n_s} |E_L|^2\right\} \tag{2.39}$$

That is, γ has a real positive value when $\omega = (\omega_L - \omega_s)$ coincides with the lattice resonance frequency ω_{LO}.

The power gain g is then given by $g = 2\ \mathrm{Re}\{\gamma\}$ because the optical power density (the light intensity) is expressed as $I = 1/2\ \varepsilon_0 n_L c |E(\omega)|^2$. At $\omega = \omega_{LO}$, the power gain is given by $g = 2\ \gamma$. The gain is independent of the instantaneous phase of the pump field, as it depends on $|E(\omega_L)|^2$. Also, the gain is in proportion to the Stokes frequency ω_s and to the inverse of the damping constant Γ.

2.3.2 Power Conversion in the Amplifying Process

Although we have so far assumed that the Stokes field is so weak that the pump field strength remains constant, the Stokes field in a high-Q resonator often becomes comparable to that of the pump field. What fraction of the pump power can be converted to the Stokes power? To answer this question, we must address the coupled equations (2.32), (2.33), and (2.34), including the change in the pump electric field strength $E(\omega_L)$. Following Bloembergen [11], this case can be solved as follows.

Instead of assuming $E \propto \exp(\gamma z)$, we assume that the amplitudes of both the pump field and the Stokes field are more general functions with respect to z:

$$E(\omega_L) = \frac{1}{\sqrt{2}} E_L(z) \exp(jk_L z - j\omega_L t) + c.c.$$
$$E(\omega_s) = \frac{1}{\sqrt{2}} E_s(z) \exp(jk_s z - j\omega_s t) + c.c. \tag{2.40}$$

Substituting these expressions into (2.32) and (2.33), we neglect the second derivative of $E_L(z)$ because $E_L(z)$ is a slowly varying function with respect to z. Then, at $\omega = \omega_{LO}$, we have:

$$\frac{\partial E_L(z)}{\partial z} = -\frac{1}{2\varepsilon_0 c^2} \frac{\omega_L^2}{k_L} \frac{\alpha_{xyz}^2}{MN\omega_{LO}\Gamma} |E_s(z)|^2 E_L(z) \tag{2.41}$$

$$\frac{\partial E_s(z)}{\partial z} = \frac{1}{2\varepsilon_0 c^2} \frac{\omega_s^2}{k_s} \frac{\alpha_{xyz}^2}{MN\omega_{LO}\Gamma} |E_L(z)|^2 E_s(z) \tag{2.42}$$

We replace $|E_L(z)|^2$ and $|E_s(z)|^2$ with the photon flux densities N_L and N_s by using the relations $N_s\hbar\omega_L \approx 1/2\varepsilon_0 n_L c|E_L(z)|^2$ and $N_L\hbar\omega_s \approx 1/2\varepsilon_0 n_s c|E_s(z)|^2$.

If we neglect spontaneous emission, we have:

$$\frac{dN_L}{dz} = -g_0 N_s N_L \tag{2.43}$$

$$\frac{dN_s}{dz} = g_0 N_L N_s \tag{2.44}$$

with

$$g_0 = \frac{2}{\varepsilon_0^2 c^2} \hbar \frac{\omega_L \omega_s}{n_L n_s} \frac{\alpha_{xyz}^2}{\omega_{LO}\Gamma MN} \tag{2.45}$$

The solution of the coupled equations is given by:

$$N_s(z) = \frac{N_s(0)(N_s(0) + N_L(0))}{N_s(0) + N_L(0)\exp\{-g_0(N_s(0) + N_L(0))z\}} \tag{2.46}$$

with

$$N_s(z) + N_L(z) = N_s(0) + N_L(0) = \text{constant} \tag{2.47}$$

The flux density $N_s(z)$ monotonically increases with z to a maximum value:

$$N_s^{\max} = N_s(0) + N_L(0) \quad \text{at} \quad z = \infty \tag{2.48}$$

This means that the whole of power from the pump beam can be converted to Stokes power. When $N_L(0) \gg N_s(0)$, the characteristic length for the power conversion, l_0, is given by:

$$l_0 = \frac{1}{g_0 N_L(0)} = \frac{1}{g} \tag{2.49}$$

That is, l_0 is equal to the inverse of the Raman gain given by (2.39).

So far, we have neglected the presence of optical loss mechanisms. There are internal absorption losses, the loss due to spontaneous Raman scattering as well as scatterings caused by inhomogeneities of refractive index in a medium, and the loss associated with reflections at the resonator mirrors. The effect of the internal losses can be taken into account as follows:

$$\frac{dN_L}{dz} = (-g_0 N_s - \alpha_L)N_L \tag{2.50}$$

$$\frac{dN_s}{dz} = (g_0 N_L - \alpha_s)N_s \tag{2.51}$$

where α_L and α_s are the power absorption coefficients for the pump and Stokes radiation. For simplicity, we assume that $\alpha_L = \alpha_s = \alpha$; then N_s can be expressed in the form:

$$N_s(z) = \frac{(N_s(0) + N_L(0))N_s(0)\exp(-\alpha z)}{N_s(0) + N_L(0)\exp\left\{g_0(N_s(0) + N_L(0))\frac{1}{\alpha}[\exp(-\alpha z) - 1]\right\}} \tag{2.52}$$

This expression is somewhat complicated, but if $\alpha z \ll 1$ holds, it reduces to a simpler form:

$$N_s(z) \simeq \frac{(N_s(0) + N_L(0))N_s(0)\exp(-\alpha z)}{N_s(0) + N_L(0)\exp\{-g_0(N_s(0) + N_L(0))z\}} \tag{2.53}$$

This expression is equal to (2.46) multiplied by $\exp(-\alpha z)$. In other words, the transmitted Stokes power is reduced only by a factor $\exp(-\alpha z)$ for $\alpha z \ll 1$.

In this discussion, we have addressed only the forward Raman scattering in which both the pump and the Stokes beams travel in the same direction. We must also consider the backward Raman scattering for the *LO* phonon mode, because in the actual Raman laser, the Stokes light, as well as the pump light, is reflected back at the resonator mirrors. The Raman gain for the backward scattering should be the same as that of the forward scattering, provided the lattice damping factor Γ can be assumed the same for both because the lattice resonance frequencies are the same. However, the power conversion feature is different from that of the forward scattering.

As illustrated in Figure 2.6, we assume that the Stokes light incident at $x = 0$ propagates to the right while the pump light incident at $x = \ell$ propagates to the left.

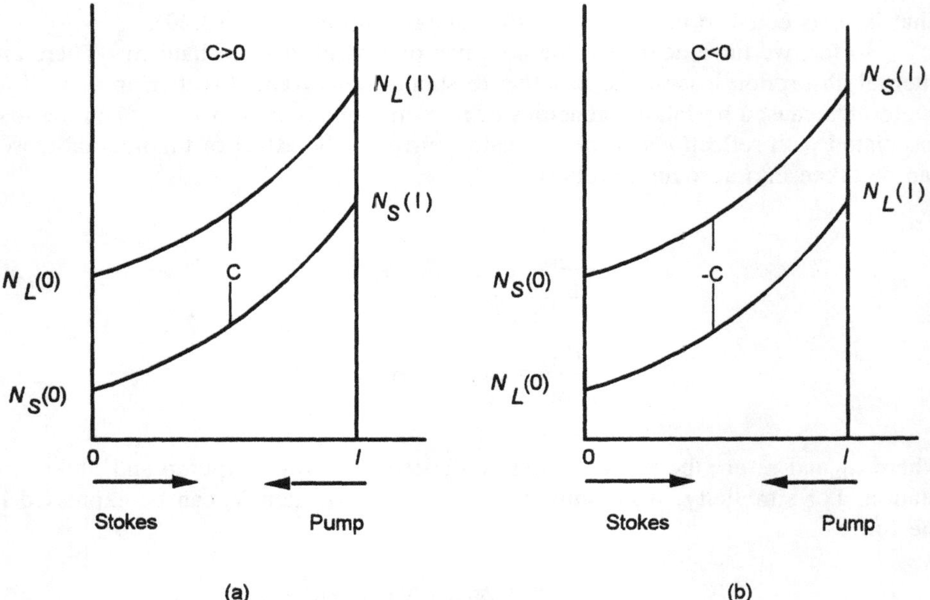

Figure 2.6 Illustration of light amplification through the backward scattering process: (a) $N_L(z) > N_s(z)$; (b) $N_L(z) < N_s(z)$.

Instead of (2.40), $E(\omega_L)$ should be written as:

$$E(\omega_L) = \frac{1}{\sqrt{2}} E_L(z) \exp(-jk_L z - j\omega_L t) + c.c. \tag{2.54}$$

with $k_L > 0$ and $k = -k_L - k_s$.

As a result, we have the following equations for N_s and N_L instead of (2.43) and (2.44):

$$\frac{dN_L}{dz} = g_0 N_s N_L \tag{2.55}$$

$$\frac{dN_s}{dz} = g_0 N_L N_s \tag{2.56}$$

The solution of these equations is given by:

$$N_s(z) = \frac{CN_s(0)}{(N_s(0) + C)\exp(-g_0 Cz) - N_s(0)} \tag{2.57}$$

with $N_L(z) - N_s(z) = C$ (constant).

The parameter C is obtained as the solution of the following equation:

$$\exp(g_0 Cl) = \frac{1}{N_s(0)N_L(l)}(N_s(0) + C)(N_L(l) - C) \tag{2.58}$$

It should be noted that C can take a negative value under some conditions; that is, $N_s(z)$ can be larger than $N_L(z)$ as illustrated in Figure 2.6(b). This result means that the whole of pump power can be converted to Stokes power even for backward Raman scattering. As will be discussed later, the Stokes field intensity in the laser resonator can exceed the pump field intensity because we usually use high-reflectance mirrors for the resonator. Therefore, the circumstances illustrated in Figure 2.6(b) often appear.

2.3.3 Effect of the Anti-Stokes Wave

So far we have neglected the presence of the anti-Stokes wave $E(\omega_a)$ with frequency $\omega_a = \omega_L + \omega_{ph}$. The coupling of the pump, Stokes, and anti-Stokes waves occurs via lattice vibrations. This was discussed in detail by Bloembergen and Shen [15].

We briefly follow their discussion. The anti-Stokes wave can produce the Stokes wave through nonlinear interaction with the $E_L^2 E_a^*$ term because we have the relation $\omega_s = 2\omega_L - \omega_a$. Similarly, the Stokes wave produces the anti-Stokes wave via the $E_L^2 E_s^*$ term because $\omega_a = 2\omega_L - \omega_s$. Referring to (2.35), we see that the following two coupled equations result after eliminating the lattice displacement $Q(\omega)$:

$$\left(\frac{\partial^2}{\partial z^2} + \frac{\omega^2}{c^2} n_s^2\right) E(\omega_s) = -\frac{1}{\varepsilon_0 c^2} \omega_s^2 [(\chi_s + \chi_{NR}) |E(\omega_L)|^2 E(\omega_s) + \{(\chi_a^* \chi_s)^{1/2} + \chi_{NR}\} E(\omega_L)^2 E^*(\omega_a)] \quad (2.59)$$

$$\left(\frac{\partial^2}{\partial z^2} + \frac{\omega^2}{c^2} n_a^2\right) E(\omega_a) = -\frac{1}{\varepsilon_0 c^2} \omega_a^2 [(\chi_a + \chi_{NR}) |E(\omega_L)|^2 E(\omega_a) + \{(\chi_s^* \chi_a)^{1/2} + \chi_{NR}\} E(\omega_L)^2 E^*(\omega_s)] \quad (2.60)$$

In these equations, χ_s and χ_a are the nonlinear susceptibility for the Stokes and anti-Stokes waves caused by the lattice vibration, while χ_{NR} has been added to express the nonresonant electronic contribution to the nonlinear susceptibility. For the longitudinal lattice vibration, χ_s and χ_a are given by:

$$\chi_s = \frac{\alpha_{xyz}^2}{MN\{\omega_{LO}^2 - (\omega_L - \omega_s)^2 + j(\omega_L - \omega_s)\Gamma\}} \quad (2.61)$$

$$\chi_a = \frac{\alpha_{xyz}^2}{MN\{\omega_{LO}^2 - (\omega_a - \omega_L)^2 - j(\omega_a - \omega_L)\Gamma\}} \quad (2.62)$$

These coupled equations can be solved under the assumption that $E(\omega_L)$ is constant. Although we do not get into the details, an interesting conclusion is that if the three waves satisfy the phase matching condition $2\mathbf{k}_L - \mathbf{k}_a - \mathbf{k}_s = 0$, there is neither amplification nor attenuation for both the Stokes and anti-Stokes waves. This can be understood as cancellation occurring between the Stokes wave and the anti-Stokes wave. In reality, however, there is a wavelength dispersion of the refractive index so that the perfect phase matching condition is usually not satisfied along the resonator axis.

If the phase mismatch is large and the waves propagating in the z directions are expressed in the forms $E(\omega_s) \propto \exp(jk_s z + \gamma_s z)$ and $E(\omega_a) \propto \exp(jk_a z + \gamma_a z)$, then γ_s and γ_a are given by:

$$\gamma_s = j\frac{1}{2\varepsilon_0 c^2}\frac{\omega_s^2}{k_s}(\chi_s + \chi_{NR})|E_L|^2 - j\left(\frac{1}{2\varepsilon_0 c^2}\frac{\omega_s^2}{k_s}\right)^2 (\chi_s + \chi_{NR})^2 |E_L|^4 \Delta k^{-1} + \cdots \quad (2.63)$$

$$\gamma_a \simeq -j\frac{1}{2\varepsilon_0 c^2}\frac{\omega_s^2}{k_s}(\chi_s + \chi_{NR})|E_L|^2 + \cdots \tag{2.64}$$

where $\Delta k = 2k_L - k_s - k_a$.

The first term in γ_s is the same as (2.39). Therefore, it is understood that the Stokes wave is amplified while the anti-Stokes wave is attenuated.

Anti-Stokes radiation via stimulated Raman scattering was first observed by Terhune and Stoicheff [16,17]. When a pump beam from a ruby laser was focussed in a benzene cell, radiation lines at frequencies $\omega_L + \omega_{ph}$, $\omega_L + 2\ \omega_{ph}$ ... were observed in a characteristic ring pattern on a film placed after the cell. The directions in which the anti-Stokes lines are observed are the ones for which the phase matching condition is satisfied. For the semiconductor Raman laser equipped with high-reflectivity mirrors, weak anti-Stokes radiation is emitted along the resonator axis, even though it is not the phase matched direction.

2.3.4 Electro-Optic Effect in the Lattice Resonance

In the previous sections, we have simply assumed that Raman scattering is caused by the nonlinearity of the permittivity with respect to the lattice displacement $Q(\omega)$. However, the longitudinal optical phonon mode and the polariton mode are accompanied by an electric field $E(\omega)$, as was discussed in Section 2.1. This electric field gives rise to a nonlinear polarization through the electro-optic effect, which is a nonlinearity of the permittivity with respect to the electric field. Therefore, the nonlinear polarization should be expressed as a sum of the electric field and lattice displacement terms, instead of (2.18). For a T_d symmetry crystal, the nonlinear polarizationcan be written in the form:

$$P_z^{NL}(\omega_s) = [d_E E_y^*(\omega) + d_Q Q_y^*(\omega)]E_x(\omega_L) \tag{2.65}$$

where x, y, and z are $\langle 100 \rangle$ axes.

Faust and Henry experimentally measured the ratio of the two contributions in GaP by mixing visible and far-infrared light [18]. The mixed light output at frequency ω_s can be related to $P_z^{NL}(\omega)$ through:

$$P_z^{NL}(\omega_s) = d(\omega_s = \omega_L - \omega)E_y^*(\omega)E_x(\omega_L) \tag{2.66}$$

where $E_x(\omega_L)$ and $E_y(\omega)$ are the electric fields of the incident visible and far-infrared laser beams. Using (2.4), we have the following expression for the nonlinear susceptibility:

$$d(\omega_s = \omega_L - \omega) = d_E\left[1 + C\left(1 - \frac{\omega^2}{\omega_0^2} - \frac{j\omega\Gamma}{\omega_0^2}\right)^{-1}\right] \tag{2.67}$$

with

$$C = (e^*/M\omega_0^2)(d_E/d_Q) \tag{2.68}$$

Figure 2.7 shows the measured frequency dependence of $d(\omega_s = \omega_L - \omega)$. It was found $C = -0.53$; that is, the electro-optic and lattice nonlinear polarizations have opposite signs in the case of GaP. The nonlinear susceptibility $d(\omega_s = \omega_L - \omega)$ approaches zero near $\hbar\omega = 250$ cm^{-1} because the two contributions cancel each other.

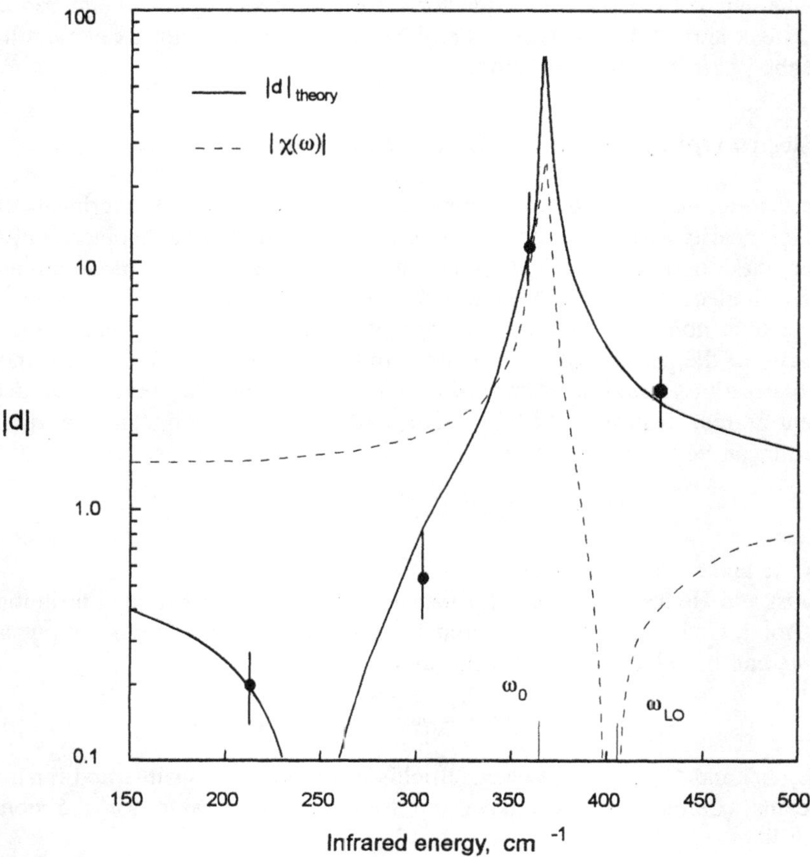

Figure 2.7 The moduli of the nonlinear and linear susceptibilities near the lattice resonance normalized to approach unity for large ω [18].

However, it should be noted that the two contributions become additive for the LO phonon mode Raman amplification because the electric field changes sign at $\omega = \omega_{TO}$. This can be seen by replacing (2.33) and (2.34) by:

$$\left(\frac{\partial^2}{\partial z^2} + \frac{\omega_s^2}{c^2} n_s^2\right) E(\omega_s) = \frac{1}{\varepsilon_0 c^2} \omega_s^2 \{d_Q Q^*(\omega) + d_E E^*(\omega)\} E(\omega_L)$$

$$= \frac{1}{\varepsilon_0 c^2} \omega_s^2 d_Q \left\{1 + \frac{1}{C}\left(1 - \frac{\omega_{LO}^2}{\omega_{TO}^2}\right)\right\} E(\omega_L) Q^*(\omega) \quad (2.69)$$

$$M(\omega_{LO}^2 - \omega^2 - j\omega\Gamma)\, Q(\omega) = \frac{d_Q}{N} E(\omega_L) E^*(\omega_s) \quad (2.70)$$

The power gain , g_{LO}, is therefore given by:

$$g_{LO} = \frac{1}{\varepsilon_0 c M N} d_Q^2 \left\{1 + \frac{1}{C}\left(1 - \frac{\omega_{LO}^2}{\omega_{TO}^2}\right)\right\} \frac{1}{\omega_{LO}\Gamma} \left\{\frac{\omega_s}{n_s} |E_L|^2\right\} \quad (2.71)$$

In the case of GaP, the power gain is larger by a factor $\{1 + 1/C(1 - \omega_{LO}^2/\omega_{TO}^2)\} = 1.41$, than that for the pure mechanical mode (TO mode).

2.4 OPTICAL FIELDS IN THE RESONATOR AND THE OUTPUT POWER

Let us consider the Raman laser, in which stimulated Raman scattering occurs in a crystal placed between the two parallel reflector mirrors forming a resonator. The Raman laser starts to oscillate when the round-trip Raman gain exceeds the total loss, which is the sum of the internal loss and the loss at the reflector mirrors. This condition can be written as:

$$R_1 R_2 \exp(g - \alpha)2l = 1 \quad (2.72)$$

where R_1 and R_2 are the reflectances of the two mirrors and α is the residual power absorption coefficient of the material. The residual absorption is usually on the order of 0.1 cm^{-1}, which is much larger than the loss due to the spontaneous Raman scattering, so that we have neglected the latter. It is interesting to note that in a Raman laser, there is not such a direct absorption as occurs in a laser diode below the lasing threshold.

The main internal loss mechanisms are free-carrier absorption and absorption due to deep levels in the forbidden band, which arises from defects and impurities in the semiconductor. These two would be neglibly small if we could make crystals of perfect quality. Another loss mechanism we should take into account is two-photon absorption. Two-photon absorption itself has little importance for the lasing threshold because the Stokes field is yet weak. However, there is a secondary effect that cannot be neglected; that is, the generation of free carriers due to the strong pump field.

Two-photon absorption occurs when the photon energy of the pump light is greater than half of the direct bandgap energy. Table 2.4 shows the two-photon absorption coefficient and the threshold wavelength of the pump light for a few semiconductors. For GaP, we need not consider two-photon absorption when $\lambda_p > 890$ nm.

From these considerations, it is seen that the only loss in the ideal case is the reflector mirror loss. We discuss the optical intensity distribution in such an ideal case. The resonator mirrors are assumed to have reflectances R_1 and R_2 for the first Stokes line, but 100% transmittances for the pump light. The pump light is introduced at $z = 0$. The wave interaction in the resonator is somewhat complicated because we must address a three-beam interaction between the pump light $E(\omega_L)$, the Stokes light transmitted in the positive direction, $E_f(\omega_s)$, which is amplified through forward scattering, and the Stokes light transmitted in the negative direction, $E_b(\omega_s)$, which is amplified through backward scattering.

Section 2.3.2 provides the basis for the following coupled equations:

$$\frac{dN_L}{dz} = -g_0 N_L N_f - g_0 N_L N_b \tag{2.73}$$

$$\frac{dN_f}{dz} = g_0 N_L N_f \tag{2.74}$$

$$\frac{dN_b}{dz} = -g_0 N_L N_b \tag{2.75}$$

First, we have the following relation:

Table 2.4
Two-Photon Absorption Coefficients

Compound	*Absorption Coefficients (cm/MW)*	*Two-Photon Energy (2 hν) (eV)*	E_{gd} *(eV)*	*Threshold Wavelength (μm)*	*Reference*
GaP	0.24	3.18	2.78	0.89	[19]
	0.21	3.56			
	0.24	3.91			
GaAs	0.06	2.34	1.5	1.65	[20]

$$N_L(z) + N_f(z) - N_b(z) = N_0 = \text{constant} \tag{2.76}$$

Although these can be solved in a general form, we only consider the limiting case when the reflectances of the mirrors are so high that the Stokes intensities become much stronger than the pump intensity. This consideration provides the maximum conversion efficiency. In such circumstances we can assume $(N_f - N_b)/N_f \ll 1$. Then, we simply have the following two coupled equations:

$$\frac{dN_L}{dz} = -g_0 N_L N_f - g_0 N_L N_b \simeq -2g_0 N_L N_f \tag{2.77}$$

$$\frac{dN_f}{dz} = g_0 N_L N_f \tag{2.78}$$

this gives

$$N_L(z) + 2N_f(z) \simeq N' = \text{constant} \tag{2.79}$$

and

$$N_L(z) = \frac{N'}{1 - \left(1 - \dfrac{N'}{N_L(0)}\right) \exp(g_0 N' z)} \tag{2.80}$$

On the other hand, at the reflector mirrors, we have the following conditions:

$$N_f(0) = R_1 N_b(0), \quad N_b(l) = R_2 N_f(l)$$

$$\text{and} \quad N_L(0) + \left(1 - \frac{1}{R_1}\right) N_f(0) = N_L(l) + (1 - R_2) N_f(l) \tag{2.81}$$

Then, it is found that the parameter N' is the solution of the following equation:

$$1 - \frac{1 - R_1 R_2}{R_1 + 1} \frac{N'}{N_L(0)} = \exp(-\, g_0 N' l) \tag{2.82}$$

When the pump power is so high that we can assume $\exp(-g_0 N' l) \ll 1$, N' approaches the limiting value:

$$N'_{limit} = \frac{1 + R_1}{1 - R_1 R_2} N_L(0) \tag{2.83}$$

Let us consider the case $R_2 \simeq 1$ and the Stokes power is taken out through the mirror with a reflectance R_1, then the limiting value of $N_f(0)$ is given by:

$$N_{f,limit}(0) = \frac{1}{2}(N'_{limit} - N_L(0)) = \frac{1}{2}\frac{R_1(1 + R_2)}{1 - R_1R_2}N_L(0) \simeq \frac{R_1}{1 - R_1}N_L(0) \quad (2.84)$$

and the output power is given by

$$N_{out,limit} = \frac{1 - R_1}{R_1}N_{f,limit}(0) \simeq N_L(0) \quad (2.85)$$

Therefore, the whole of the pump power can be converted to output power, even when the output mirror has a high reflectance, such as $R_1 \simeq 90\%$.

Figure 2.8 illustrates the relation between the pump and the Stokes intensities in the resonator and the output power. It should be noted that the internal Stokes intensity given by (2.84) can be much higher than the pump intensity when the reflectance R_1 becomes high. For example, when $R_1 = 0.9$, the internal Stokes intensity should be almost one order of magnitude higher than the pump intensity.

However, if we cannot neglect the internal losses, it is obvious that such a high intensity does not build up in the resonator. Also, we have neglected generation of

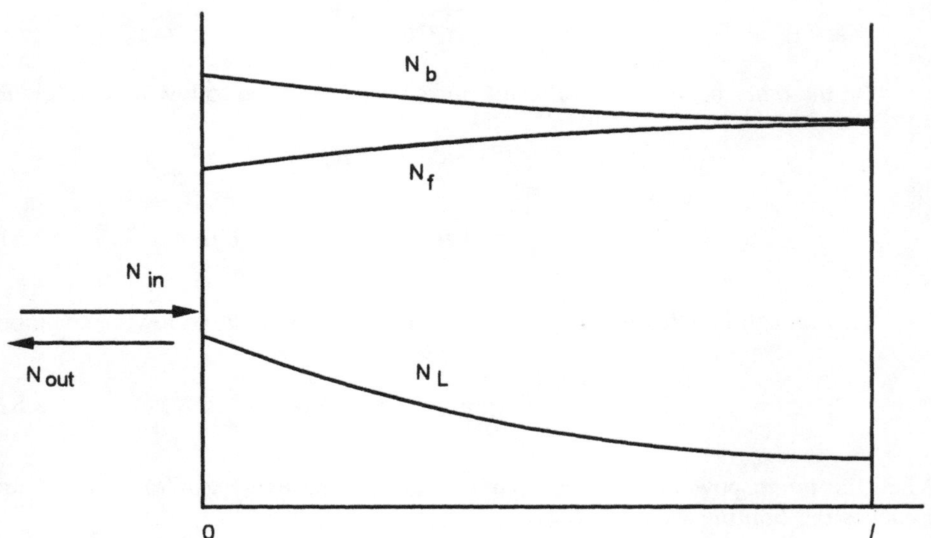

Figure 2.8 The spatial dependencies of the Stokes and pump powers in a resonator when the mirrors have high reflectances at the Stokes wavelength.

higher order Stokes lines. Once the first Stokes light intensity becomes higher than the pump intensity, this first Stokes light can act as a pump light source to generate oscillation of the wave at frequency $\omega_L - 2\ \omega_{ph}$; that is, the second Stokes light wave. Then, the second Stokes wave excites the third Stokes wave and so on. Such a cascade lasing phenomenon is generally observed in various kinds of Raman lasers. We will describe this in the semiconductor Raman laser in Chapter 3. Cascade lasing can be suppressed if the resonator mirrors have low reflectance at the second Stokes frequency.

2.5 RAMAN AMPLIFICATION IN OPTICAL FIBER

Raman amplification and oscillation in optical fibers were first demonstrated by Stolen, Ippen, and Tynes [21]. Figure 2.9 illustrates their Raman laser experimental system. Although the Raman gains in glass materials are very small compared to those in crystals and liquids, significant Raman amplification can be obtained by long-length propagation in low-loss optical fiber.

The Raman gain bands in crystals or liquids are narrow, and they are homogeneously broadened with a Lorenzion shape:

$$g(\omega) = g_{\max} \frac{1}{1 + (\omega - \omega_{LO})^2/\Gamma^2} \tag{2.86}$$

Conversely, the Raman gain bands in glasses are widely spread as shown in Figure 2.10. They are inhomogeneously broadened because of the distribution of local potentials that occurs in glass materials.

The Raman gain of a single-mode fiber with a core diameter of 3.8 μm and a length of 590 cm was measured by an experimental arrangement similar to that shown in Figure 2.9 but without resonator mirrors [22]. The 526-nm line from a pulsed Xe laser with power-level 2 W to 24 W was used as a pump source. For a signal light

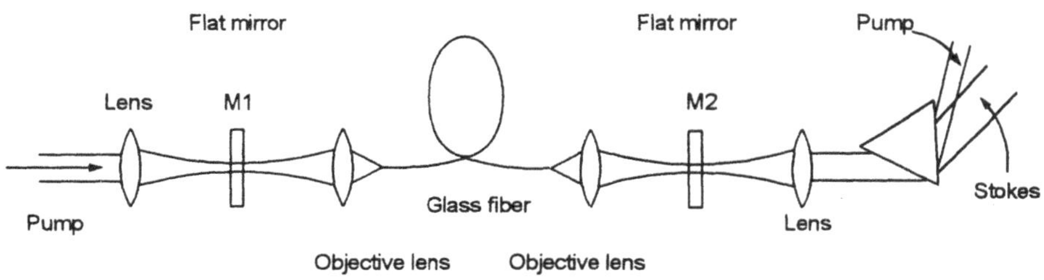

Figure 2.9 Experimental arrangement for the observation of fiber Raman oscillation [21].

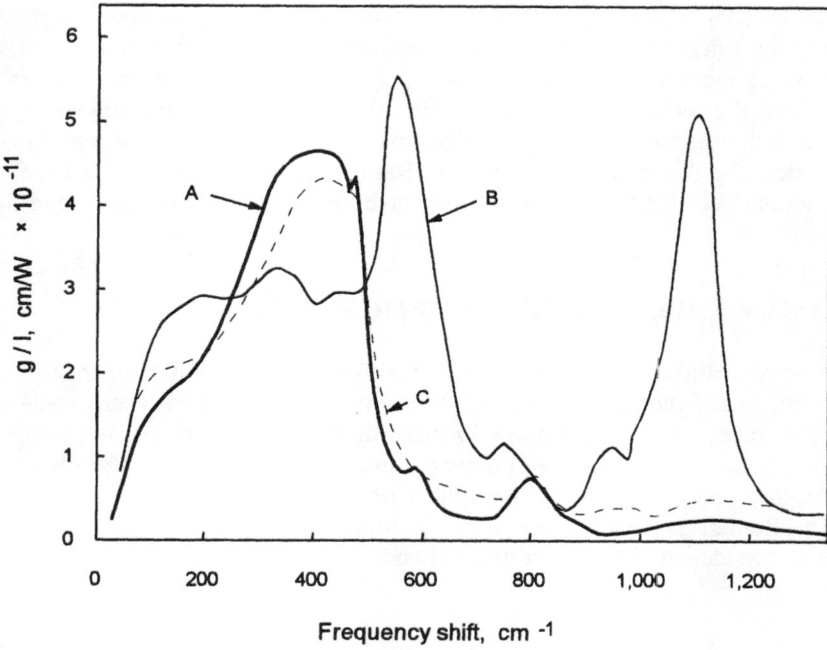

Figure 2.10 Raman gain curves for three representative glasses: (A) fused quartz; (B) soda-lime-silicate (20:10:70); (C) pyrex [21].

to be amplified, the 535.3-nm line from the Xe laser was used. The latter has a frequency shift of 330 cm^{-1} from the pump light so that it is in the vicinity of the maximum gain position, as depicted in Figure 2.10. As a result, the Raman gain coefficient, g_0, can be obtained, which is the gain for a unit of pump power density:

$$g_0 = 1.5 \times 10^{-11} \text{ cm/W}$$

This gain coefficient is about two orders of magnitude smaller than those in semiconductors.

In comparison with the Raman gain, the gain for stimulated Brillouin scattering in silica fiber is much higher. The Brillouin gain coefficient for fused quartz was measured to be approximately $4{,}3 \times 10^{-9}$ cm/W, with a Brillouin shift $\Delta\upsilon = 32.2$ GHz and a bandwidth of less than 100 MHz [23]. Actually, amplification via stimulated Brillouin scattering was suppressed because the linewidths of the pump and signal lights were broader than the bandwidth of the stimulated Brillouin scattering.

The optical fiber Raman amplifier can be used for light amplification at signal frequencies for which no suitable rare-earth doped fiber amplifiers are available.

Another advantage of the fiber Raman amplifier is in the soliton propagation. Narrow optical pulses in the soliton mode can propagate in ideal form for a long distance only when the medium has no propagation loss. The very small optical loss that occurs in fibers can be compensated by the Raman amplification process. More details of soliton propagation will be discussed in Section 7.1.2.

REFERENCES

[1] M. Born and K. Huang, *Dynamical Theory of Crystal Lattices*, London: Oxford University Press, 1954.

[2] M. Hass, "Lattice Reflection," Semiconductors and Semimetals, Vol. 3, *Optical Properties of III-V Compounds*, ed. R.K. Willardson and A.C. Beer, New York: Academic Press, 1967, pp. 3–16.

[3] D.A. Kleinman and W.G. Spitzer, "Infrared Lattice Absorption of GaP," *Phy. Rev.*, Vol. 118, 1960, pp. 110–117.

[4] M. Balkanski, "Photon-Phonon Interactions in Solids," *Optical Properties of Solids*, ed. F. Abelès, Amsterdam: North-Holland, 1972, pp. 529–651.

[5] W.D. Johnston, Jr. and I.P. Kaminow , "Temperature Dependence of Raman and Rayleigh Scattering in $LiNbO_3$ and $LiTaO_3$," *Phys. Rev.*, Vol. 168, 1968, pp. 1045–1054.

[6] J.G. Skinner and W.G. Nilsen, "Absolute Raman Scattering Cross-Section Measurement of the 992 cm^{-1} Line of Benzene," *J. Opt. Soc. Am.*, Vol. 58, 1968, pp. 113–119.

[7] N. Bloembergen, G. Bret, P. Lallemand, and P. Simova, "Controlled Stimulated Raman Amplification and Oscillation in Hydrogen Gas," *IEEE J. Quant. Electron.*, Vol. 3, 1967, pp. 197–201.

[8] R.R. Loudon, "The Raman Effects in Crystals," *Advances in Physics*, Vol. 13, 1964, pp. 423–482.

[9] J. Nishizawa and K. Suto, "Semiconductor Raman and Brillouin Lasers for Far-Infrared Generation," *Infrared and Millimeter Waves*, Vol. 7, ed. K. J. Button, New York: Academic Press, 1983, pp. 301–320.

[10] M. Cardona, M.H. Grimsditch, and D. Olego, "Theoretical and Experimental Determination of Raman Scattering Cross Sections in Simple Solids," *Light Scattering in Solids*, ed. Birman, New York: Plenum Press, 1979, pp. 249–255.

[11] C. Flytzanis, "Infrared Dispersion of Second-Order Electric Susceptibilities in Semiconducting Compounds," *Phy. Rev.*, Vol. 6, 1972, pp. 1264–1277.

[12] V. S. Ryazanov, V. Garelik, G.V. Peregudow, M.M. Sushchinskii, and V.A. Chirkou, "Cross Sections for the Raman Scattering of Light in GaP and CdS Crystals," *Sov. Physics-Solid State*, Vol. 10, 1968, pp. 1508–1509.

[13] S. Ushioda, A. Pinczuk, W. Taylor and E. Burstein, "The Raman Scattering Intensities of Zincblende Type Crystals," *II–VI Semiconducting Compounds*, ed. D.G. Thomas, New York: W. A. Benjamin, 1967, pp. 1185–1203.

[14] N. Bloembergen and Y.R. Shoen, "Coupling between Vibration and Light Waves in Raman Laser Media," *Phys. Rev. Lett.*, Vol. 12, 1964, pp. 504–507.

[15] N. Bloembergen, *Nonlinear Optics*, New York: W. A. Benjamin, 1965.

[16] R.W. Terhune, *Bull. Am. Phys. Soc.*, Vol. 8, 1963, p. 359.

[17] B.P. Stoicheff, *Phys. Rev. Lett.*, Vol. 7, 1963, p. 186.

[18] W.L. Faust and C.H. Henry, "Mixing of Visible and Near-Resonance Infrared Light in GaP," *Phys. Rev.*, Vol. 17, 1966, pp. 1265–1268.

[19] C.B. de Araujo and H. Lotem, "New Measurements of the Two-Photon Absorption in GaP, CdS, and ZnSe Relative to Raman Cross Sections," *Phys. Rev.*, Vol. 18, 1978, pp. 30–38.

[20] A. Azema, J. Botineau, F. Gires and A. Saïssy, "Guided-Wave Measurement of the 1.06 mm Two-Photon Absorption Coefficient in GaAs Epitaxial Layers," *J. Appl. Phys.*, Vol. 49, 1978, pp. 24–28.

[21] R.H. Stolen, E.P. Ippen and A.R. Tynes, "Raman Oscillation in Glass Optical Waveguide," *Appl. Phys. Lett.*, Vol. 20, 1972, pp. 62–64.

[22] R.H. Stolen and E.P. Ippen, "Raman Gain in Glass Optical Waveguides," *Appl. Phys. Lett.*, Vol. 22, 1973, pp. 276–278.

[23] E.P. Ippen and R.H. Stolen, "Stimulated Brillouin Scattering in Optical Fibers," *Appl. Phys. Lett.*, Vol. 21, 1972, pp. 539–541.

Chapter 3
Bulk Semiconductor Raman Laser

3.1 FUNDAMENTAL CHARACTERISTICS OF THE SEMICONDUCTOR RAMAN LASER

3.1.1 Lasing Experiment of the Bulk Semiconductor Raman Laser

Among various semiconductors, GaP is the most suitable Raman laser material because it has a large Raman polarizability, small lattice damping constant, and is highly transparent in the optical frequency region below its absorption edge. The bandgap energy of GaP is 2.2 eV at room temperature, which corresponds to a wavelength of 550 nm or, in other words, an optical frequency of 545 THz. GaP is an indirect bandgap material, so residual absorption due to the perturbation of the absorption edge by impurities, as is observed in direct bandgap materials such as GaAs and InP, is scarcely observed.

Most of the first stimulated Raman scattering experiments used the Q-switched high-power ruby laser with wavelength $\lambda = 690$ nm, which was focussed onto samples without any resonator structure. Although the Raman gain is proportional to the pump frequency, there is a possibility of a larger absorption loss when the pump frequency comes near the absorption edge of the semiconductor. Therefore, for pumping the semiconductor Raman laser, a more suitable pump source is the Nd: YAG laser with wavelength $\lambda = 1.064$ μm. (The optical frequency is 281.7 THz.)

The first demonstration of the semiconductor Raman laser was performed using the system illustrated in Figure 3.1 [1]. We used undoped n-type GaP bulk crystals grown by the liquid encapsulated Czochralsky method with carrier concentrations $n < 3 \times 10^{16}$ cm^{-3}. A GaP crystal was cut to a rectangular shape approximately 10 to 15 mm long and 5 by 5 mm in the cross section, and both end-faces were polished optically flat and parallel to each other to within 1.6×10^{-4} rad. The two end-faces were antireflection coated with SiO films to reduce the reflection loss. The measured transmission loss for a single path was less than 10%. The crystal was placed between two plane dielectric mirrors with reflectances $R \simeq 99\%$, which formed a Fabry-Perot resonator.

The crystal and the resonator were mounted in a cryostat at a temperature of approximately 100 K. It was later found that although the threshold pump power was lower at 100 K than at room temperature, threshold pump power did not so strongly depend on the temperature as was generally observed in laser diodes. The pump beam from a Q-switched YAG laser with a diameter approximately 6 mm was introduced into the crystal without a focussing lens, with a small incident angle of approximately 1° to 2° from the resonator axis.

Oscillation modes depend on the crystal direction to the resonator axis, as is expected from the spontaneous Raman-scattering spectra in different crystal directions (Figure 2.4). An *LO* phonon mode oscillates when the crystal direction parallel to the resonator axis is $\langle 100 \rangle$, as shown in Figure 3.2(a).

On the other hand, for the oscillation of a *TO* phonon mode, the crystal direction parallel to the resonator axis must be a little off $\langle 110 \rangle$, as will be discussed later. (See Figure 3.2(b).) Actually, we used a 7°-off crystal to obtain oscillation of the *TO* phonon mode. Both kinds of *TO* phonon modes are observed. One is the

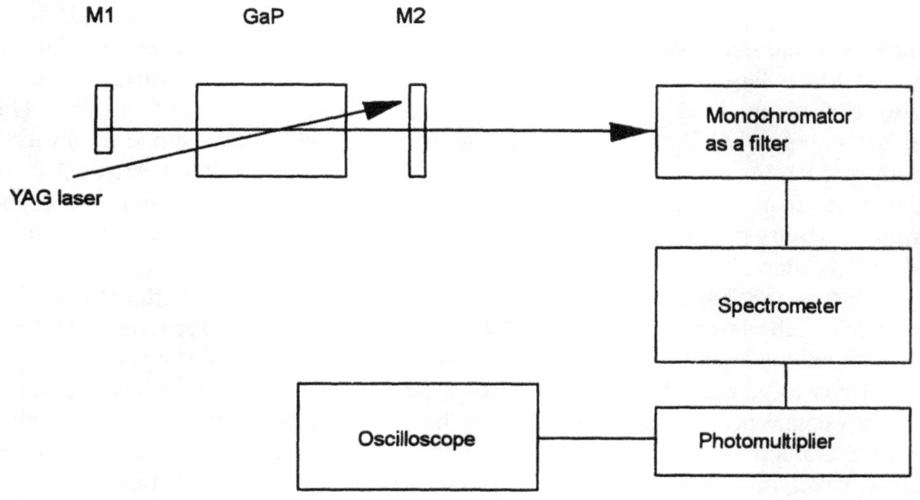

Figure 3.1 Experimental setup for the semiconductor Raman laser [1].

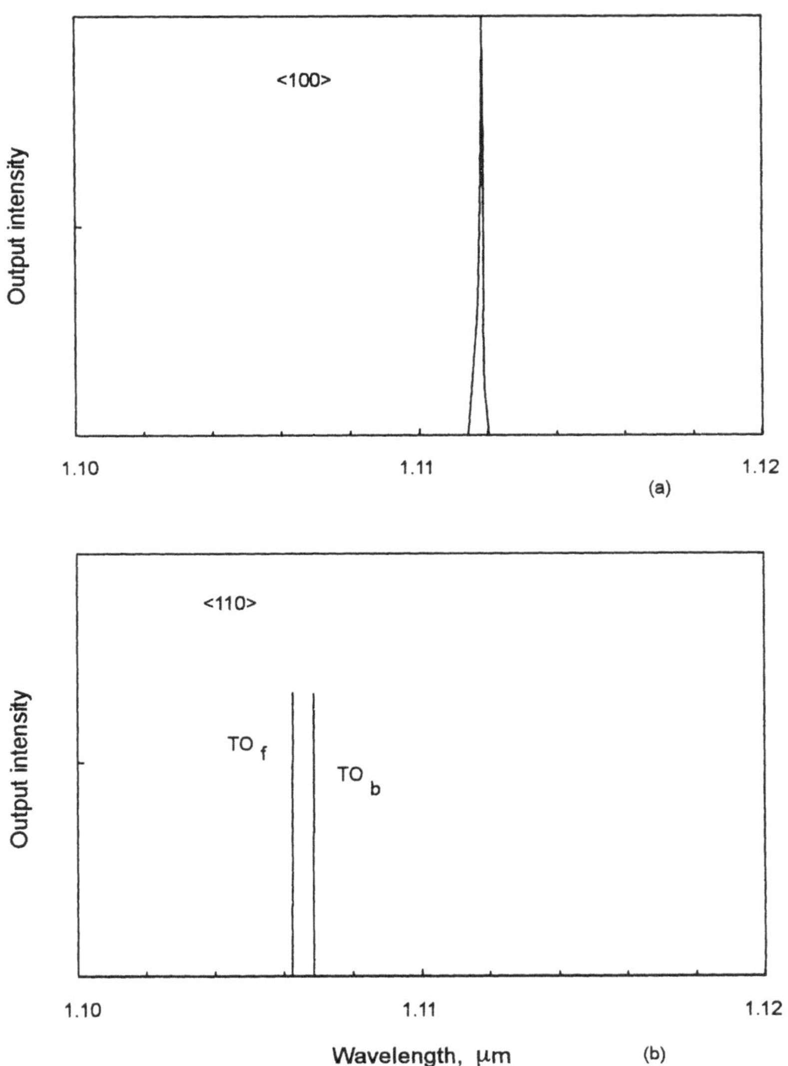

Figure 3.2 The semiconductor Raman laser oscillations for the (a) ⟨100⟩ and (b) ⟨110⟩ crystal directions parallel to the resonator axis [1].

polariton mode arising from forward scattering, designated as TO_f, while the other, designated as TO_b, is the pure transverse mode arising from backward scattering. The pump thresholds for the LO and TO_f modes are compared in Figure 3.3.

Because there is very little dispersion in the LO phonon mode, the LO phonon oscillation occurs at a relatively low threshold pump power and with a constant frequency shift. The constant frequency shift is a very desirable characteristic for frequency selection in a heterodyne demodulation system for near-infrared optical waves [2]. On the other hand, the TO_f phonon mode is useful as a far-infrared radiation source and also as a frequency-tunable near-infrared laser, although the threshold pump power is relatively high.

We point out that unless the crystal direction and incident angle are properly selected, usually only the LO phonon mode will oscillate because it has a much lower threshold. In the following discussion, we will clarify the oscillation conditions for the two different modes as a function of crystal direction and incident angle of the pump radiation relative to the resonator axis.

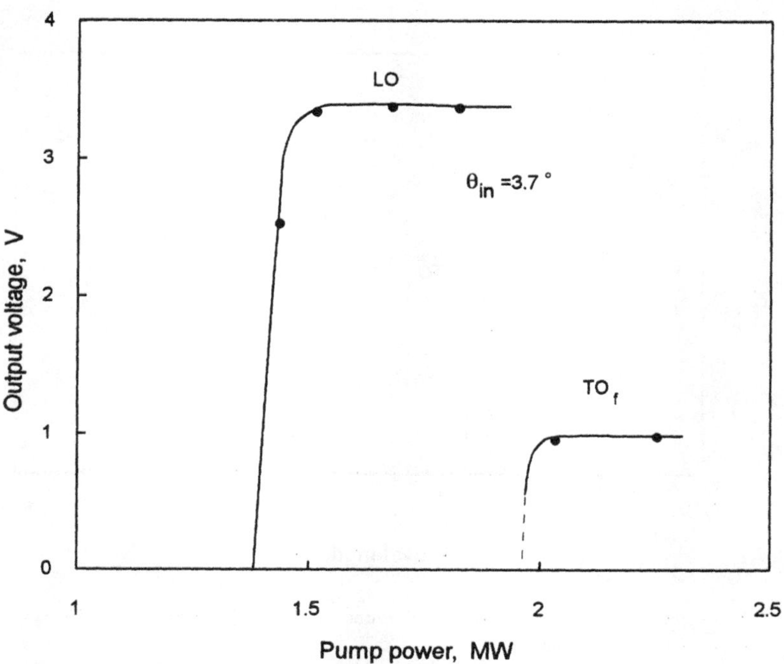

Figure 3.3 Lasing thresholds for the LO and TO_f mode oscillations [1].

Angular Dependences [3]

Raman scattering is related to the derivative of optical polarizability per unit volume $\alpha_{ijk} = \partial\alpha_{ij}/\partial Q_k$ where Q_k means the relative displacement of the two sublattices. As was discussed in 2.2, the Raman-scattering efficiency is written as:

$$S = A(e_L^i \alpha_{ijk} \xi_k e_s^j)^2 \tag{3.1}$$

where $\mathbf{e_L}$ and $\mathbf{e_s}$ indicate the unit vectors along the polarization directions of the incident and scattered photons, respectively, and $\boldsymbol{\xi}$ is the polarization direction of phonons.

Figure 3.4 illustrates the relation between the directions of the three waves and the crystal axes.

For simplicity, the resonator axis is chosen in the (001) plane, and the polarization of the incident light is in the $\langle 001 \rangle$ direction. Let θ be the angle between the resonator axis and the $\langle 010 \rangle$ crystal direction. Taking into account that the only nonzero components are $\alpha_{xyz} = \alpha_{yzx} = \alpha_{zxy}$ in a zinc-blende crystal, then (3.1) can be expressed as:

$$S = A\alpha_{x,y,z}^2[(\xi_y \cos\theta + \xi_x \sin\theta)^2]. \tag{3.2}$$

where the x and y axes are chosen to be the $\langle 100 \rangle$ and $\langle 010 \rangle$ crystal directions. For *LO* phonons, the phonon wave vector $\mathbf{q}$ and its polarization $\boldsymbol{\xi}$ are parallel. Therefore, the part of (3.2) inside the parentheses becomes:

$$f^{LO} = \cos^2(2\theta + \theta_{IS}), \tag{3.3}$$

where θ_{IS} is the angle between the wave vectors of the scattered photons and phonons, and f^{LO} is called the angular factor for the *LO* phonon mode. From the phase matching condition $\mathbf{k_L} = \mathbf{k_s} + \mathbf{q}$, the relation between θ_{IS} and θ_{in} is given by:

$$\sin\theta_{IS} = \frac{k_L}{q}\sin\theta_{in} = \frac{n_L\omega_L}{cq}\sin\theta_{in} \tag{3.4}$$

where $\mathbf{k_L}$ and $\mathbf{k_s}$ are the wave vectors of the incident and Stokes light, respectively, and n_L is the refractive index at the pump light frequency ω_L.

Similarly, the angular factor for the *TO* phonon mode polariton becomes:

$$f^{TO} = \sin^2(2\theta + \theta_{IS}), \tag{3.5}$$

because $\mathbf{q}$ is perpendicular to $\boldsymbol{\xi}$. If θ_{in}, the pump beam direction inside the crystal, is too large, homogeneous pumping along the resonator axis becomes difficult.

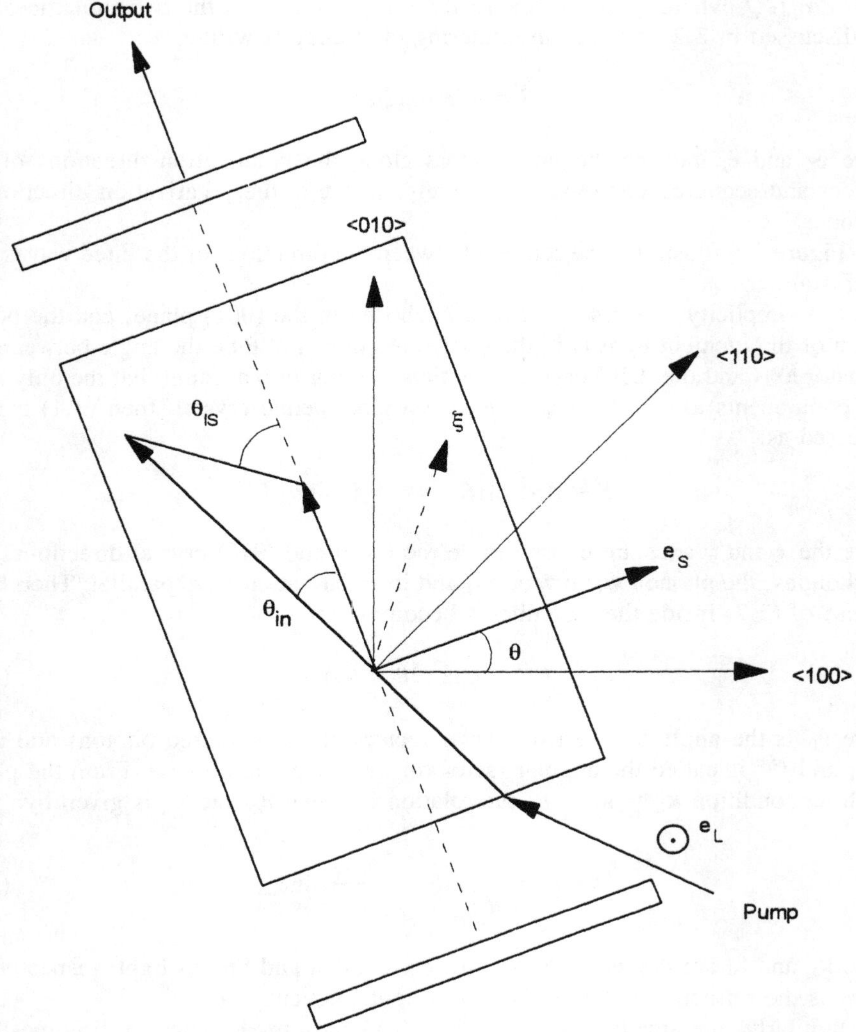

Figure 3.4 Relationship between crystal directions, polarizations, and wave vectors [3].

Whereas, if θ_{in} is too small, the component of the pure *TO* phonon in the polariton mode becomes small, and the scattering efficiency reduces, as will be shown later. In general, a suitable range of θ_{in} is 0.5° to 5° for the *TO* mode.

LO Phonon Mode. For oscillation in the *LO* phonon mode, the crystal direction along the resonator axis must be close to the ⟨100⟩ direction. From the relation given by (3.3), the angular factor for the *LO* phonon mode, f^{LO}, can be calculated as a function of the incident angle of the pump light, θ_{in}, for different crystal directions. Figure 3.5 shows the three cases where θ, the angular deviation from ⟨100⟩ axis, is 0°, 7.5°, or 15°.

It is seen that the maximum scattering efficiency is obtained when the crystal direction parallel to the resonator axis is slightly misoriented from the exact ⟨100⟩ axis when the pump beam is introduced at a finite incidence angle. In the case of the *LO* mode, the effect of the angular factor can be directly confirmed by measurement of the lasing threshold because there is no change in gain factor like that occurring in the *TO* mode. Figure 3.6 shows that when $\theta = 0°$, the threshold input power at $\theta_{in} = 2.6°$ is approximately twice of that at $\theta_{in} = 1.35°$, which agrees with the calculation in Figure 3.5.

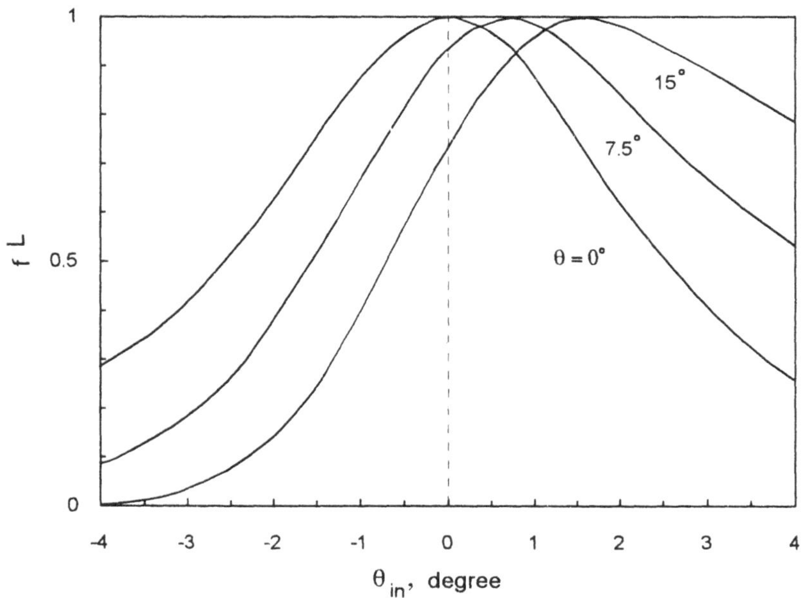

Figure 3.5 Calculated angular factors of GaP for 1.064-μm pumping [3].

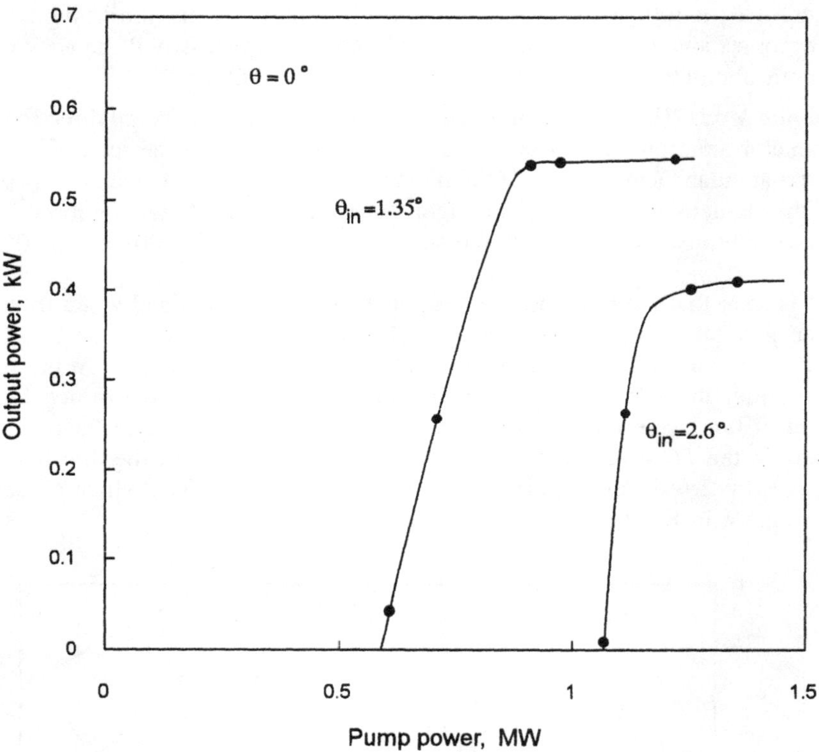

Figure 3.6 Relationship between input and output power of the semiconductor Raman laser for different incident angles [3].

TO Phonon Mode. For the forward scattering *TO* phonon mode (i.e., polariton mode), we must consider the dispersion relation discussed in Section 2.1, which is expressed as:

$$\frac{c^2}{\omega^2}q^2 = \frac{\varepsilon_\infty}{\varepsilon_0} + \frac{\omega_p^2}{\omega_0^2 - \omega^2 - j\omega\Gamma}, \quad \omega_p^2 = \frac{Ne_B^{*2}}{\varepsilon_0 M} \tag{3.6}$$

where ω is the frequency of the polariton mode phonon that varies with wave vector $\mathbf{q}$, N is the number of unit cells in a unit volume, e_B^* the effective charge of an anion and cation, M the reduced mass of the two ions, ω_0 is the pure transverse phonon frequency, and Γ the damping constant of the vibration.

The frequency ω and the wave vector $\mathbf{q}$ of the polariton mode that interacts with the pump light can be determined by the wave vector matching condition $\mathbf{k}_L =$

$\mathbf{k}_s + \mathbf{q}$, which can be written in the form:

$$c^2q^2 = \left(n_L + \omega_L \frac{dn_L}{d\omega_L}\right)^2 \omega^2 + c^2k_Lk_s \cos^2\theta_{in} \tag{3.7}$$

where $dn_L/d\omega_L$ is the frequency dispersion of the refractive index at the pump frequency ω_L.

These two relations are calculated in Figure 3.7 (a, b) for two incident wavelengths (6,328 Å and 1.064 μm). Each figure also shows θ_{IS}, which is calculated by (3.4).

To measure the angular factor for the *TO* mode, we prepared three different crystal samples, for which the ⟨110⟩ crystal direction relative to the resonator axis were $\varphi = 0°$, 7°, or 22° (referring to Figure 3.4, $\varphi = 45 - \theta$). Figure 3.8 shows the angular factor for these three crystals, calculated from (3.3) using the θ_{IS} given in Figure 3.7.

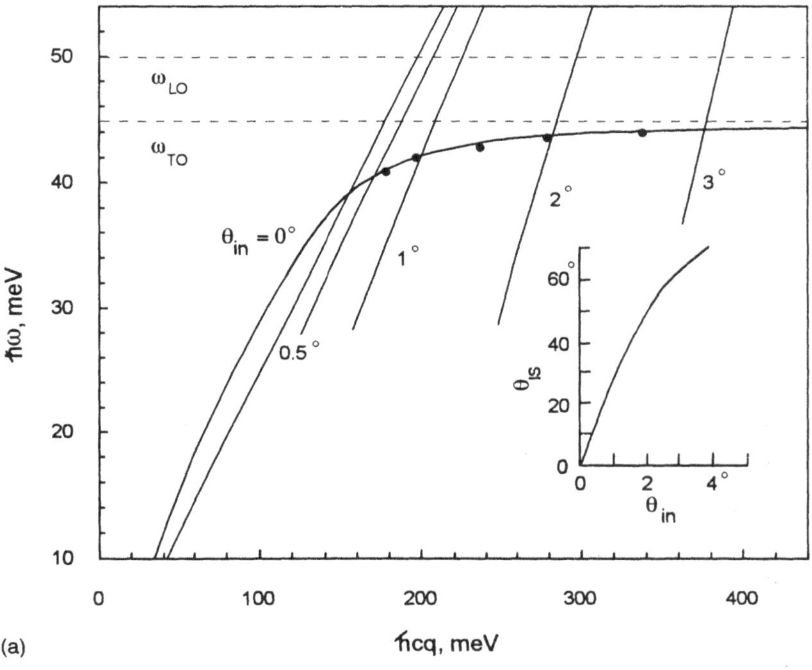

Figure 3.7 Dispersion curve and incident angle θ_{in} for GaP: (a) for 6,328-Å pumping (the points are measured by spontaneous scattering); (b) for 1.064-μm pumping (the points are where lasing has been observed) [3].

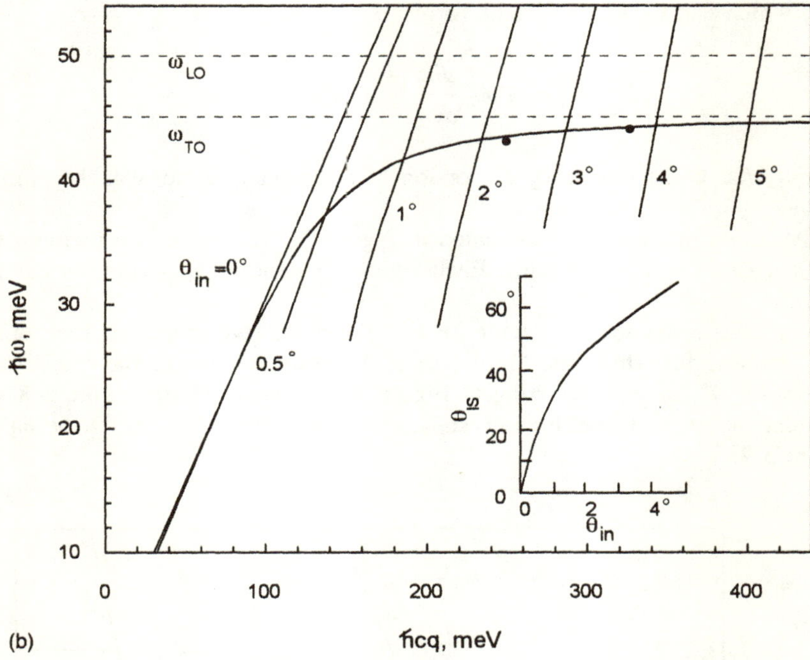

Figure 3.7 (Continued)

The angular factor of the crystal with $\varphi = 22°$ is higher than 80% over a wide range of θ_{in} ($0.5 - 6°$ at 1.064 μm). Whereas in the exact ⟨110⟩ crystal, the angular factor is considerably reduced at θ_{in} larger than 1°. Other than this factor, it should be noted that if θ_{in} becomes small, the scattering efficiency will be reduced because the component of the pure *TO* phonon in the polariton mode reduces, as is discussed in Section 3.5.

Detailed experimental analysis of the angular dependence of the scattering efficiency or Raman gain was made by measuring spontaneous scattering of He-Ne laser radiation. Although the angular dependences are slightly different for the two wavelengths (6,328 Å and 1.064 μm) because of dispersion of the refractive index, the spontaneous-scattering experiment at 6,328 Å is very useful for predicting the angular dependence at 1.064 μm.

Figure 3.9 shows the measured intensities of the spontaneous scattering by TO_f phonons at a constant incident angle $\theta_{in} = 2.75°$ for the crystal directions $\varphi = 0$, ± 7, and $\pm 22°$. The results agree well with the calculation in Figure 3.8.

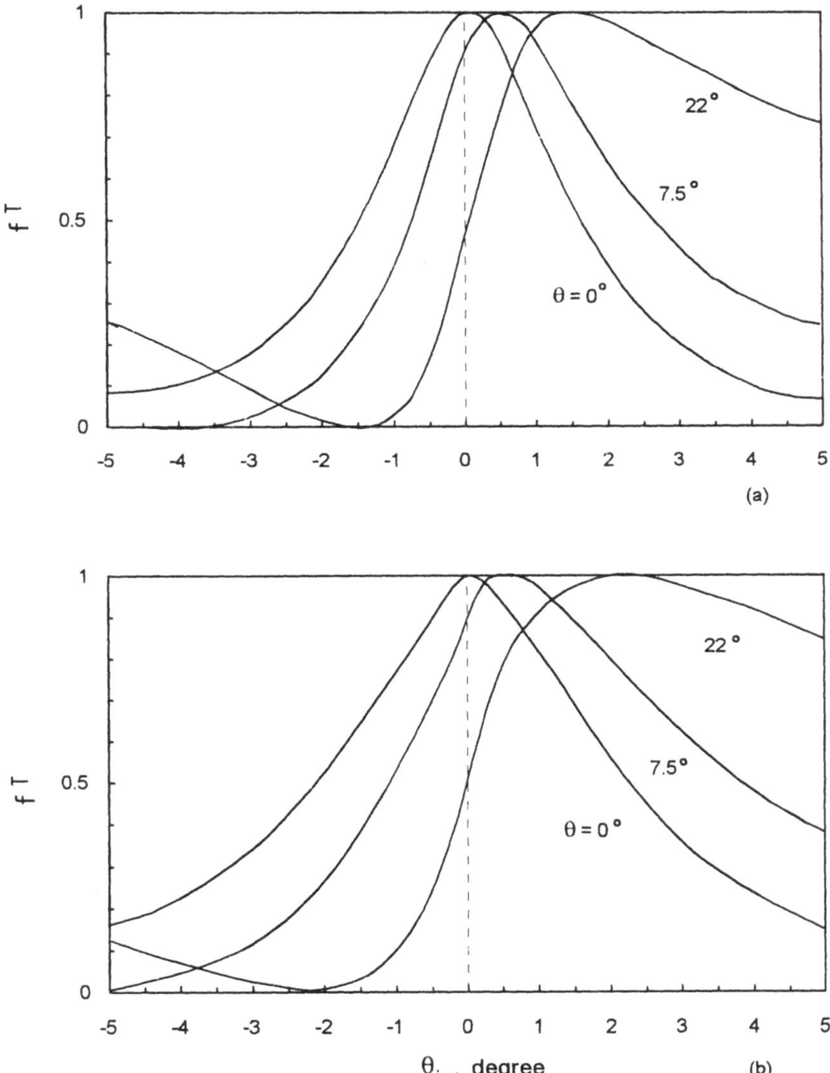

Figure 3.8 Calculated angular factors of GaP: (a) for 6,328-Å pumping; (b) for 1.064-mm pumping ($\phi = \pi/4 - \theta$; i.e., the off-angle from the ⟨110⟩ directions) [3].

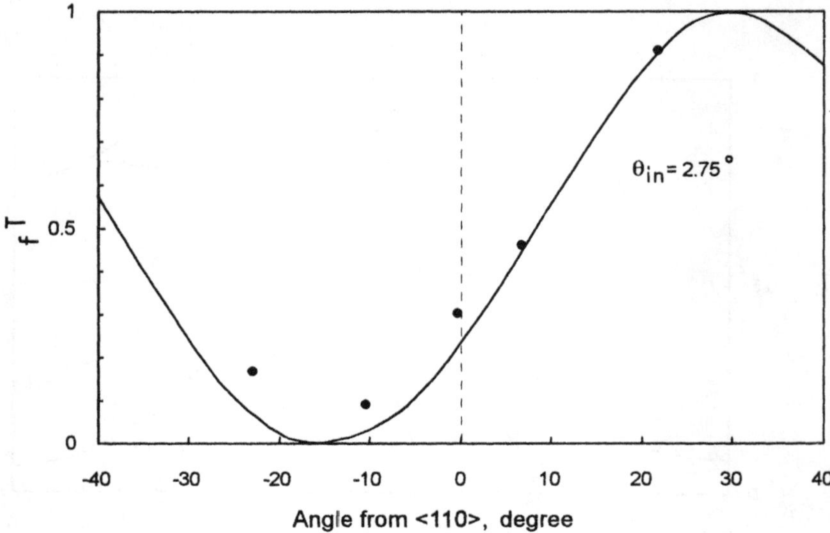

Figure 3.9 Intensity of spontaneous *TO* phonon scattering as a function of the off-angle ϕ from the $\langle 110 \rangle$ direction ($\theta_{in} = 2.75°$) [3].

However, the angular factor as a function of incident angle θ_{in} cannot be obtained directly from the *TO* phonon scattering intensity. This is because in the polariton mode, the scattering efficiency or Raman gain changes not only with the angular factor but also with the coupling ratio between mechanical vibration and the electromagnetic wave (see Section 3.5). However, we can estimate the angular factor of the *TO* phonons, f^T, by measuring that of the *LO* phonons, f^L, then applying the relation $f^T \approx 1 - f^L$, unless θ_{in} is so small that θ_{IS} for *LO* phonons departs considerably from that for *TO* phonons.

Measured *LO* phonon intensities include a component of backward scattering due to reflection of incident and scattered radiation at the crystal end surfaces. The angular factor is therefore obtained from the relation $f^L = (I - I_{\min})/(I_{\max} - I_{\min})$, where $I_{\min}$ is the minimum intensity of the scattered radiation as a function of θ_{in}.

Figure 3.10 shows the result for the two cases $\varphi = 0$ and $\varphi = 22°$. The small discrepancies observed at the smaller angle may be due to the finite cone angle used for detection of the scattered radiation.

3.1.2 Characteristics of the LO Phonon Mode Oscillation

We briefly describe how the lasing characteristics of the *LO* phonon mode change when the crystal length, temperature, or mirror reflectances are varied [3].

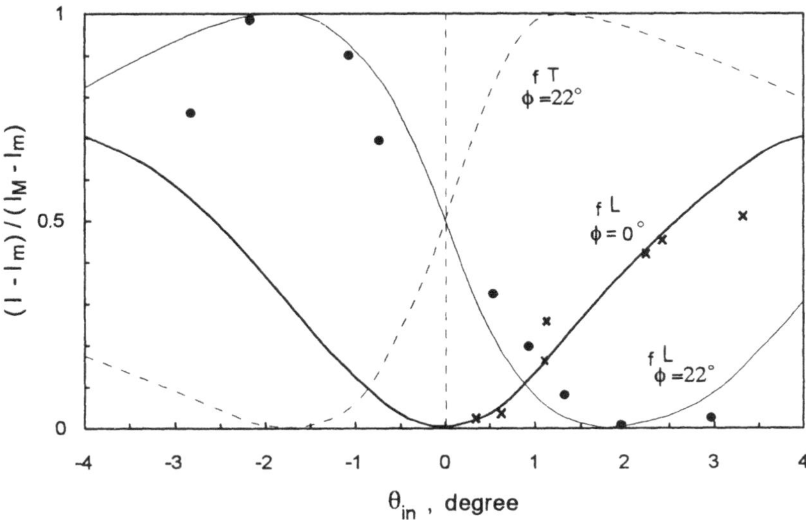

Figure 3.10 Intensities of spontaneous *LO* phonon scattering as a function of the incident angle θ_{in} for $\phi = 0$ and 22° [3].

Figure 3.11 shows the lasing threshold and the output power for three different crystal lengths when 98% mirrors are used and the temperature is 100 K. Lasing of the *LO* phonon mode is observed even in the 3.8-mm sample. Use of a short crystal is important for an optical integrated circuit, such as in the integration of a semiconductor Raman or Brillouin laser with an injection laser as pumping source.

The fact that the threshold pump power reduces with increasing crystal length means that the internal absorption loss is less than the losses due to mirror reflection and diffraction in free space. We will discuss the results of internal loss measurements in Section 3.2 and the Raman gain measurements in Section 7.3.

It is very easy to make the *LO* phonon oscillate at low temperature. Once the *LO* phonon mode ($\omega_L - \omega_{LO}$) oscillates, this, in turn, excites the ($\omega_L - 2\omega_{LO}$) mode. When the gain is high enough, the latter mode also oscillates. Therefore, cascade lasing oscillations occur if high-reflectivity mirrors with wide wavelength band are used. Figure 3.12 shows that three or four cascade oscillations have occurred [3]. This may be useful for an infrared source covering some wavelength ranges, but if we want to obtain only the first order *LO* phonon mode oscillation, it is necessary to depress the higher order *LO* phonon mode by making the mirror reflectance-band narrow.

Figure 3.13 compares the Stokes output at room temperature and 100 K for a crystal 13-mm long when the reflectivity of the output mirror is 90% [4]. The threshold input optical power at room temperature is about 3.5 times the 100 K threshold.

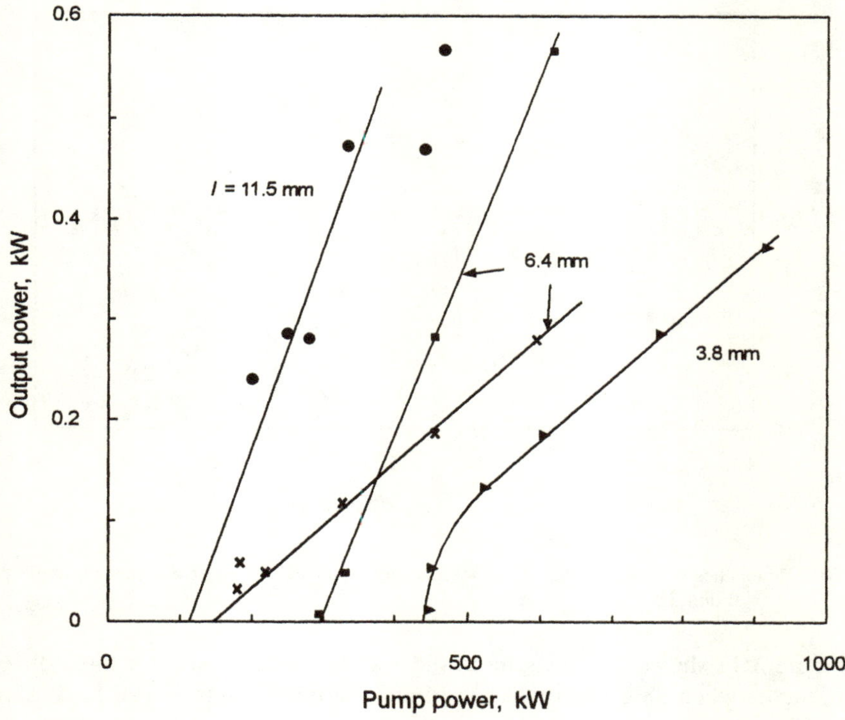

Figure 3.11 Relationships between the input power and output power of the semiconductor Raman laser for different crystal lengths. Spot diameter = 2 mm, R = 98%, θ_{in} = 1.3°, temperature = 100 K [3].

The conversion efficiency of the Stokes radiation from the pump radiation is about 1.7% at 100 K. As shown in Figure 3.13(a), oscillation is observed at room temperature even with a 70% reflection mirror, and the maximum output power exceeds 40 kW.

It is understood that the observed increase of the threshold with temperature is not so severe as in the case of laser diodes. The temperature dependence of the threshold is believed to be caused by the increase in the free-carrier concentration and the increase in the lattice damping constant.

Figure 3.14 shows experimental data for a shorter crystal, 3.8-mm long, at room temperature [4]. The threshold power increases slightly, but stable oscillation is possible. By increasing the reflectance of the output mirror to 99.5%, the oscil-

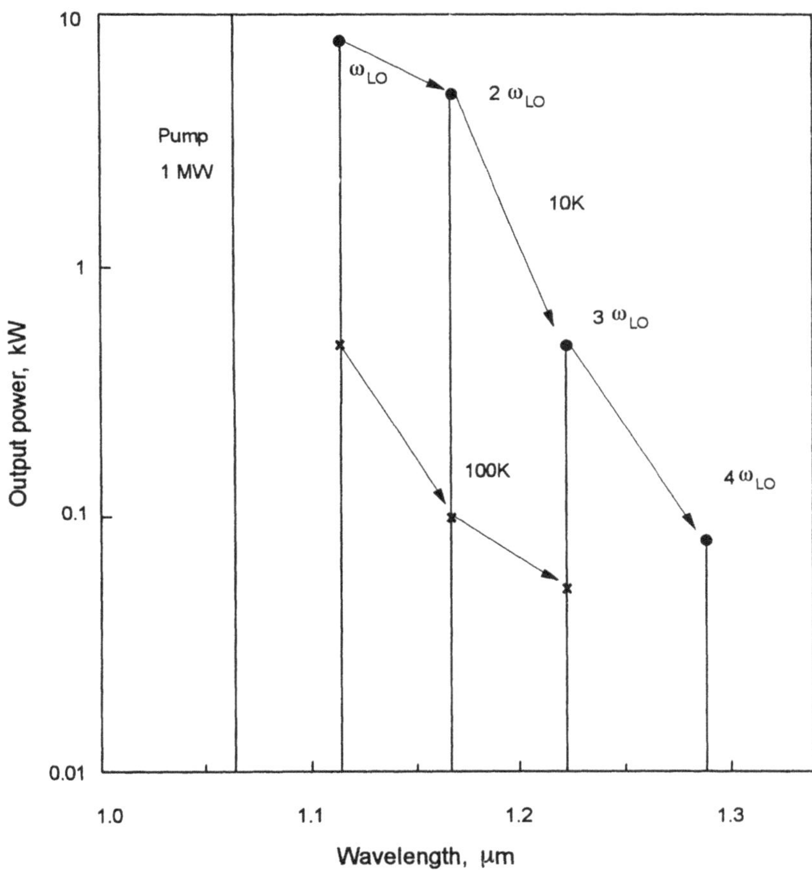

Figure 3.12 Cascade lasing of the *LO* phonon mode at 10 K and 100 K [3].

lation is easily observed without causing surface damage, and the threshold optical power density is 6.8×10^6 W/cm^2. This level of power density is lower than the optical power density of a normal double-heterostructure injection laser. For example, 10 mW output power from a double-heterostructure laser with an active layer cross section of 0.1 by 1 μm corresponds to 10^7 W/cm^2. This fact means that the injection laser is a very promising pump source for a semiconductor Raman laser. In particular, integration of an injection laser and a semiconductor Raman laser would provide a new coherent radiation source and also an optical heterodyne demodulator.

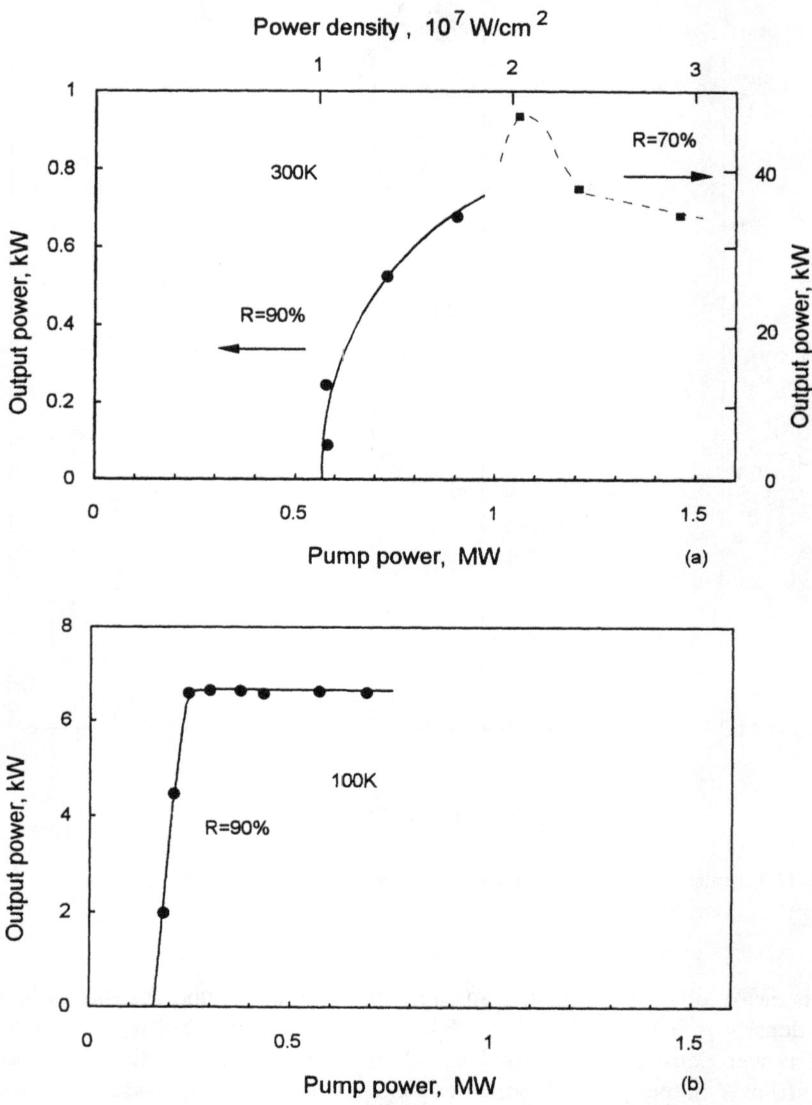

Figure 3.13 Stokes output power as a function of the incident power at room temperature (a) and 100 K (b) for a crystal of 13-μm in length. The diameter of the incident beam is 2.5 mm. R is the reflectivity of the output mirror [4].

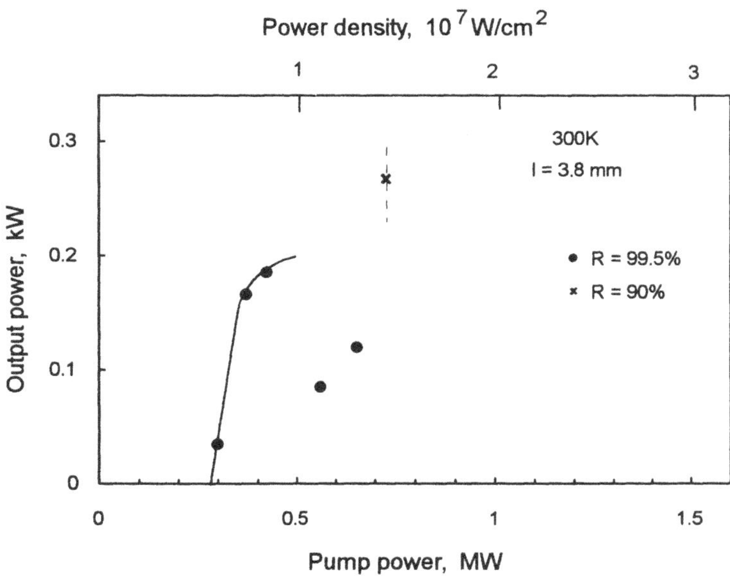

Figure 3.14 Stokes output power as a function of the incident power at room temperature for a crystal of 3.8 mm in length. The diameter of the incident beam is 2.5 mm [4].

3.1.3 Methods of Pump Beam Introduction

In the previous experiment, the cross section of the crystal is much larger than that of the incident beam so that the optical resonator is not bound in the lateral directions.

As the next step, we have made the thickness of the crystal very thin to reduce the incident optical power. As shown in Figure 3.15(a), the length of the resonator is made as short as the crystal length [4]. The threshold power density is expected to become smaller because the resonator is entirely filled by the active medium, and the rise time of the Stokes light becomes shorter. Both the upper and lower surfaces of the crystal have been optically polished to guide the beams inside the crystal. The pump beam is incident on the crystal at the edge of the mirror M_1, and multiple reflections occur in the resonator.

The cross section of the beam that comes into the resonator is given by $2Lt \sin \theta_{in}$, where L and t are the length and thickness of the crystal, respectively.

Another method of incidence uses a cubic polarizer as shown in Figure 3.15(b) [4]. The polarization direction of the incident beam $\mathbf{e}_L$ is parallel to the $\langle 010 \rangle$ direction and that of the Stokes light parallel to $\langle 100 \rangle$. The length of the polarizer should be made as short as possible to reduce the diffraction loss. We have examined two

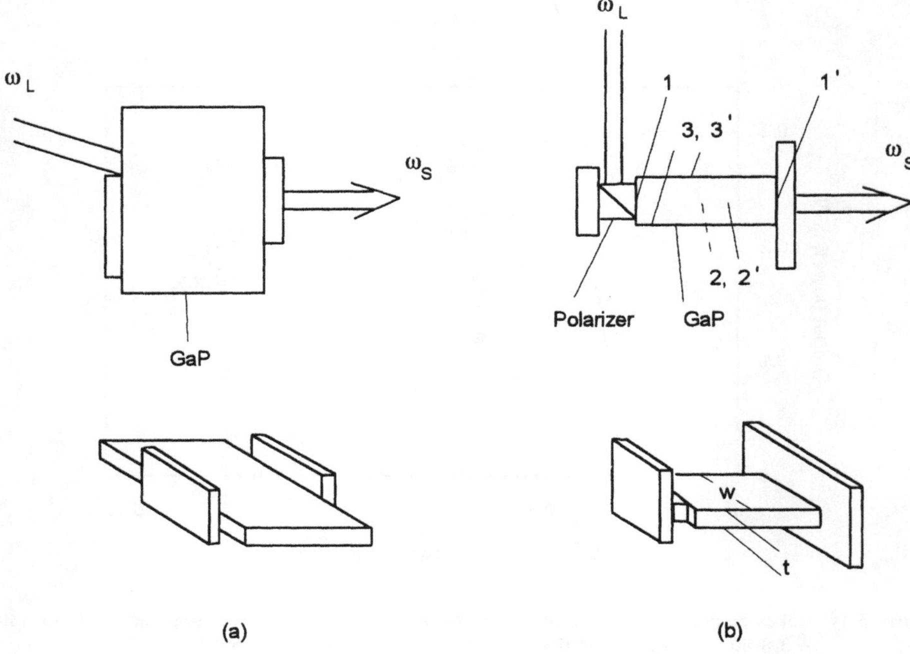

Figure 3.15 Constructions of the semiconductor Raman laser using a thin platelet GaP crystal: (a) the pump beam is incident on the crystal at the edge of the mirror with a small angle; (b) a polarizer is used.

polarizers, one 1.8 mm in length and the other 1.0 mm in length. The end-faces of the polarizer are antireflection coated to give no reflection loss for the Stokes radiation.

The Raman laser crystal has a rectangular shape with all its side walls optically polished to give a waveguiding characteristic for both the Stokes and incident optical beams owing to the high refractive index of GaP ($n = 3.2$ at 1.1 μm). The fabrication process is as follows: A thin plate is cut from a single crystal GaP with an electron concentration of approximately 2×10^{16} cm^{-3}. The two end-faces (1 and 1′ in Figure 3.15(b)) of the plate are optically polished so that they are parallel within 0.3 m rad, and the two wide surfaces (2 and 2′ in Figure 3.15(b)) are optically polished so that the deviation from an exact right angle to the first set of the faces is within 2 m rad. The plate is etched with a solution with H_2O_2:HNO_3 = 1:1 for 2 min; the etched thickness is approximately 2 μm. Strong optical absorption occurs at the polished surface without etching. Therefore, the etching process is effective both in increasing the threshold for the surface damage formation by the incident optical beam and in decreasing the absorption loss of the laser. After the antireflection coating is made on the first set of the planes, the plate is finally cut to a bar, and the remaining two surfaces (3 and 3′ in Figure 3.15(b)) are optically polished so that the deviation from a right angle to the first and second sets of the planes is within 3m rad. These sets of surfaces are not etched.

The reflectivity of the resonator mirrors is approximately 99% to 99.5%, chosen to be as high as possible, and end-faces of the crystals (1 and 1′) and the polarizers are antireflection coated with a transparency higher than 99%. The important losses are, therefore, due to absorption in the crystal and diffraction occurring at the end-face of the crystal facing the polarizer, which has no waveguiding effect.

Figure 3.16 compares the two methods of introducing the pump beam for a crystal with thickness 400 μm and length 5 mm. For small angle incidence, θ_{in} has been fixed at 2.5°. On the other hand, for the incidence through the polarizer, the length of the polarizer is 1.8 mm.

The threshold power density is lowest for incidence through the polarizer (4×10^6 W/cm^2), which can be attributed to the fact that the angular factor discussed

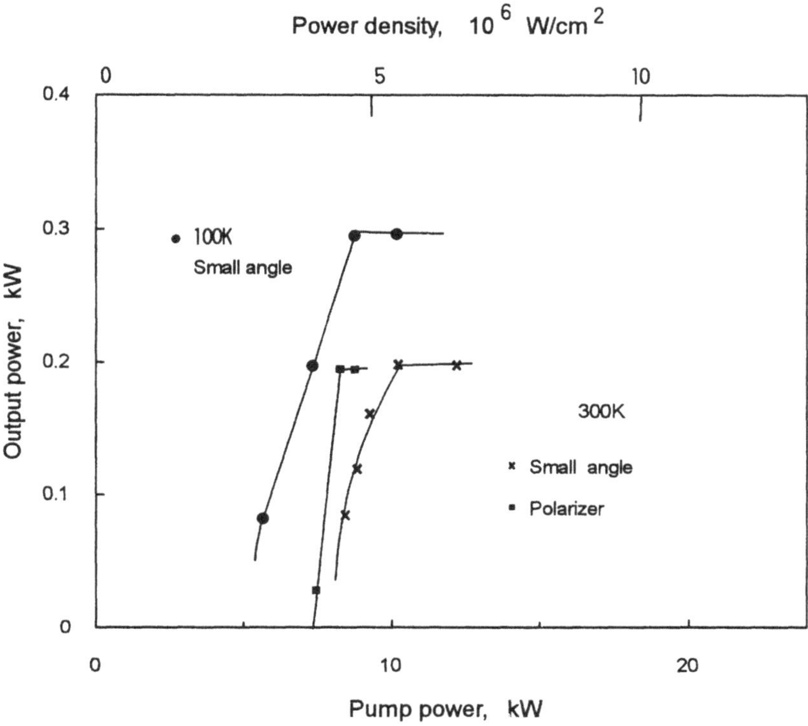

Figure 3.16 Stokes output powers of Raman lasers with and without a polarizer [4].

in Section 3.1.1 is at a maximum because the pump and Stokes radiations are collinear in the crystal.

It is obvious that a waveguiding structure with a smaller cross section will further decrease the threshold optical power considerably.

We have actually examined thinner bulk crystals. The threshold optical input power densities for 160- and 380-μm thick crystal are compared in Figure 3.17(a,b). If a 1.0-mm cubic polarizer is used, the increase in the threshold input power density in the 160-μm thick crystal is slight, not serious; 1.7×10^6 W/cm^2 for the 380-μm thick crystal and 2.1×10^6 W/cm^2 for the 160-μm crystal.

These results mean that the diffraction losses at the crystal end-face facing the polarizer are yet unimportant. However, if the thickness is further reduced, this loss will become more important.

In contrast to the thickness t, the width w (the direction of which is in the plane of incidence of the polarizer as shown in Figure 3.15(b)) affects the threshold more severely.

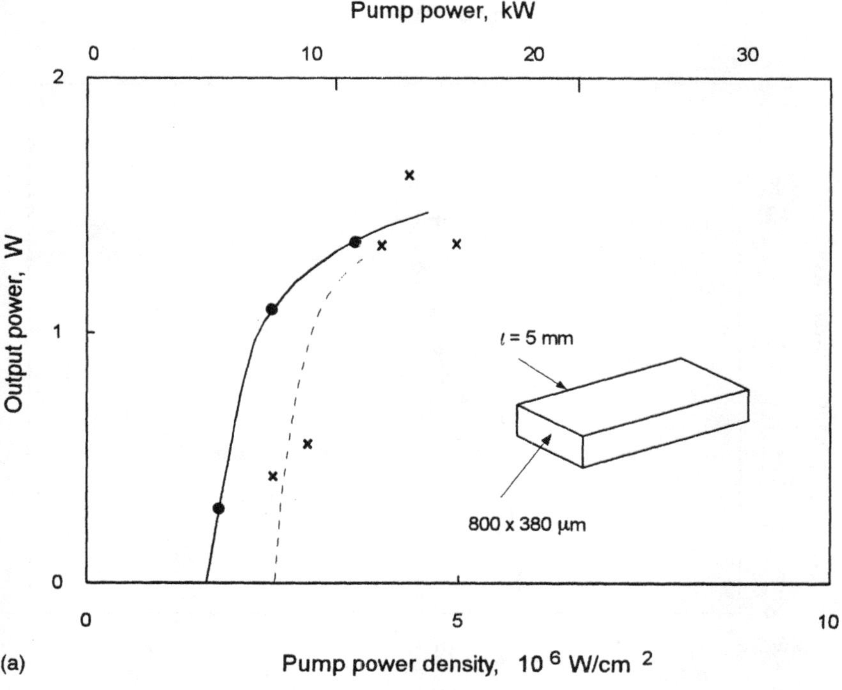

Figure 3.17 Stokes output power as a function of incident optical power. (a) t = 380 μm. (b) t = 150 μm. ● = 1.0-mm polarizer; × = 1.8-mm polarizer [5].

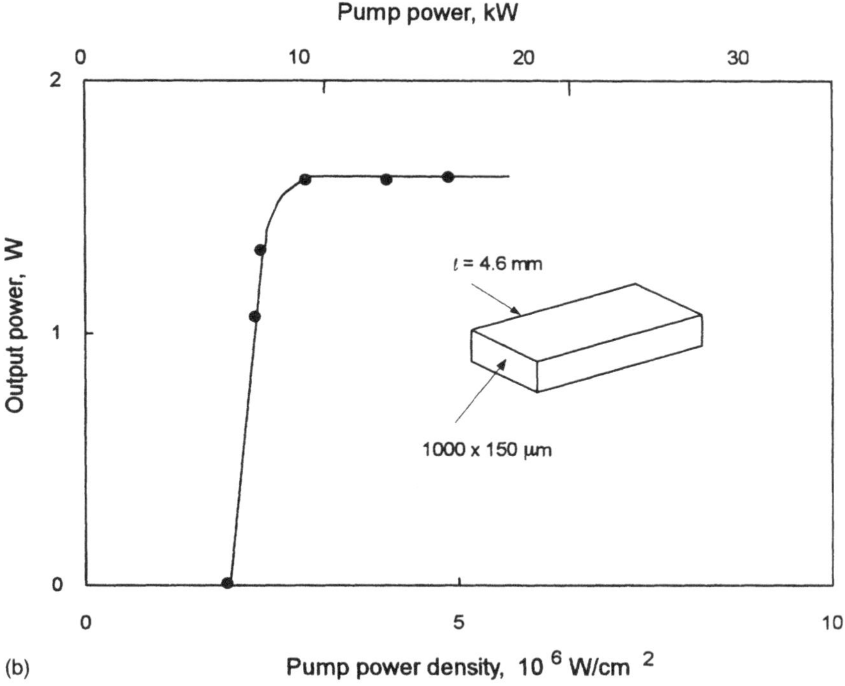

Figure 3.17 (Continued)

For a thin and wide structure, as shown in Figure 3.17, there is little diffraction loss in the crystal region because of waveguiding inside the crystal [5]. The diffraction loss due to the incorporation of the polarizer can be roughly estimated as follows, by assuming that the optical field in the crystal region is confined in the crystal. We consider only one-dimensional Fresnel diffraction takes place, on the assumption that $t \ll W$. The Fresnel number is given by $N = t^2/\lambda 2\ell_p$, where λ is the wavelength in the polarizer of which the refractive index is 1.5 and ℓ_p is the length of the polarizer.

The field intensity returning by reflection at the resonator mirror outside the slit of the crystal end-face is given by the well known Fresnel integrals A_c and A_s [6], on the assumption that the slit is very wide compared with the wavelength.

$$I(\nu)/I_0 = \frac{1}{2}\left[\left\{\frac{1}{2} - A_c(\nu)\right\}^2 + \left\{\frac{1}{2} - A_s(\nu)\right\}^2\right]$$

$$A_c(\nu) = \int_0^\nu \cos\left(\frac{\pi}{2}\tau^2\right) d\tau$$

$$\nu = -\sqrt{\frac{2}{\lambda 2 l_p}}\, x$$

$$A_s(\nu) = \int_0^\nu \sin\left(\frac{\pi}{2}\tau^2\right) d\tau \tag{3.8}$$

where x is the position from one of the edges of the slit. This gives the integrated intensity outside the slit as $2\int_0^\infty I(\nu)\,dx \simeq 0.3\sqrt{\lambda l_p} I_0$. The loss L_D is given by dividing by tI_0; thus, $L_D \simeq 0.3\sqrt{\lambda l_p}/t \simeq 0.2/\sqrt{N}$. This corresponds to the loss for the lowest order mode.

In the case of the crystal with thickness 400 μm, L_D is approximately 2% to 2.6%. These values are smaller than the absorption loss in the crystal (14% 10%). Therefore, the difference in the threshold values for the two polarizers may be due to misalignment of the components, which would cause a larger effect for the longer polarizer.

On the other hand, in the case of the crystal thickness of 160 μm, L_D is about 5%, comparable to the absorption loss. This will therefore result in a small increase of the threshold optical input power.

3.2 INTERNAL LOSS MECHANISMS IN THE GaP RAMAN LASER

The fundamental absorption edge of GaP is 2.2 eV or 550 nm in wavelength at room temperature. The transparent wavelength range of high-purity GaP extends from below the absorption edge to near the edge of the lattice reflection band, where an absorption band associated with higher order multiphonon transitions appears. The latter causes significant absorption in the wavelength region beyond 12 μm, as shown in Figure 3.18 [7].

However, in the near infrared region, the main loss mechanisms are free-carrier absorption and absorption via deep levels. The free-carrier absorption coefficient α_f is thought to be in proportion to the square of the light wavelength λ, and is given by:

$$\alpha_f = \frac{Nq^2\lambda^2}{8\pi^2 n_\lambda m^* \tau_c c^3} \tag{3.9}$$

where N is the free-carrier concentration, m^* is the effective mass of free carriers, τ_c is the mean time between collisions, and n_λ the refractive index.

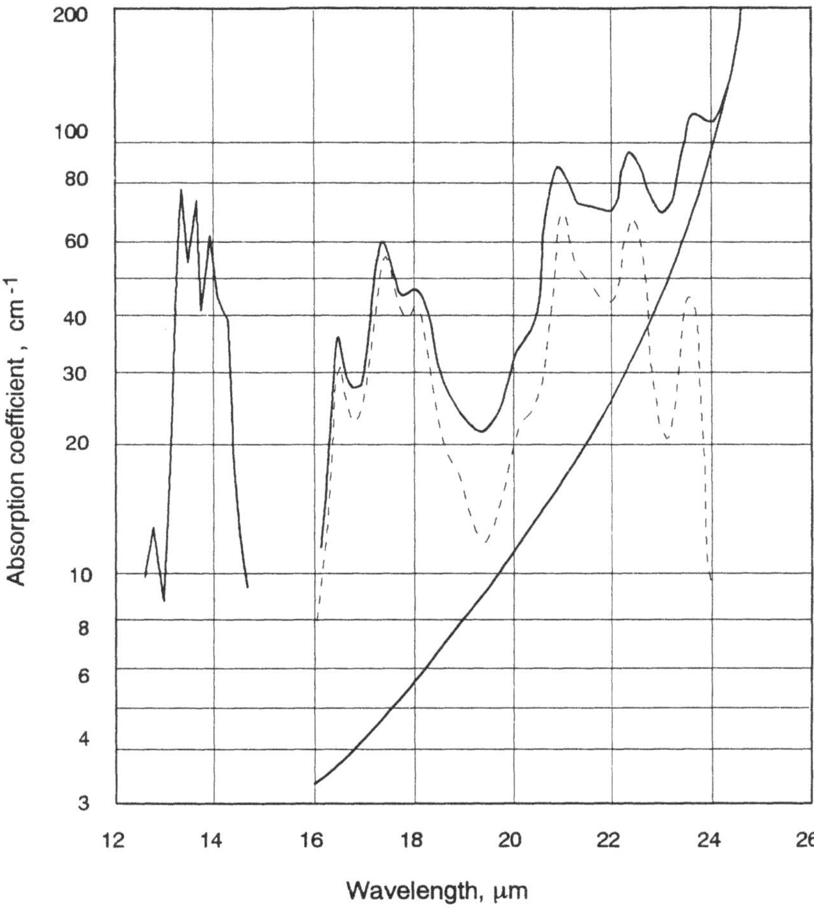

Figure 3.18 The experimental absorption coefficient α, the absorption coefficient for the fundamental resonance α_f, and the absorption coefficient for the combination bands $\alpha_c = \alpha - \alpha_f$ [7].

It should be noted that there is also an interband transition for n-type GaP between the X and X' Brillouin zones as shown in Figure 3.19 [8]. Free-carrier absorption in the classical meaning is small at lower temperatures, but the absorption due to the interband transition is thought to be basically independent of temperature because electrons can be excited from the donor levels as long as they are shallow levels retaining the X zone wave function character. Actually, it is not known which is the dominant absorption mechanism at room temperature in the near infrared wavelength region.

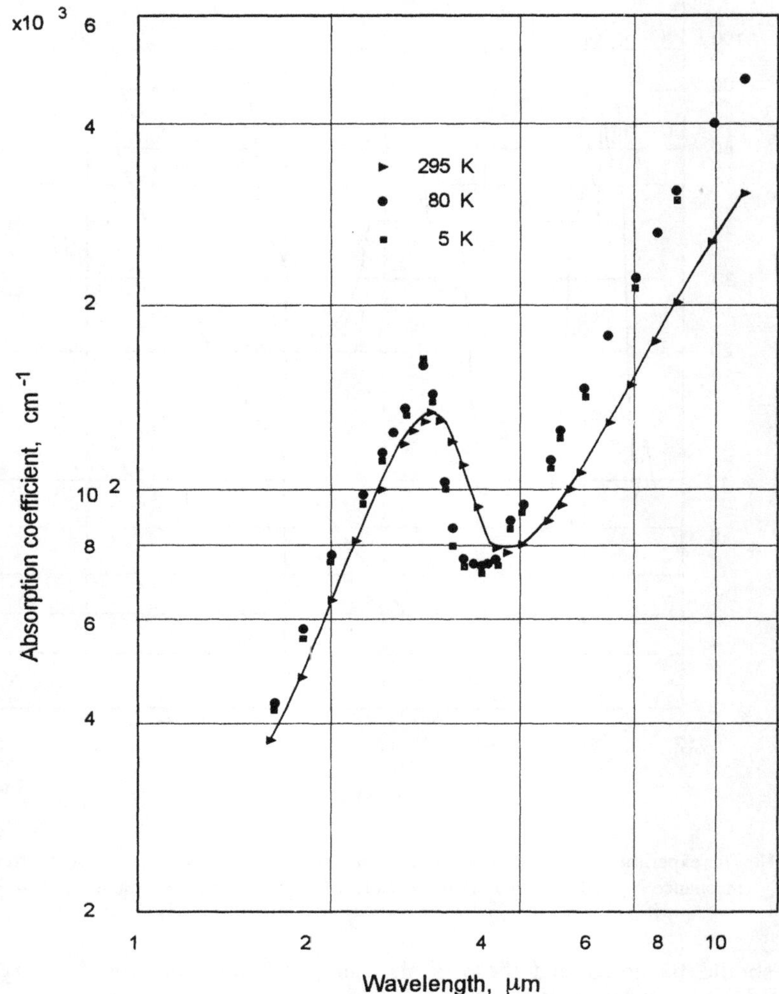

Figure 3.19 The absorption spectrum of *n*-type GaP ($N = 1 \times 10^{18}$ cm^{-3}). The rapid rise of α at long wavelength is due to free-carrier absorption [8].

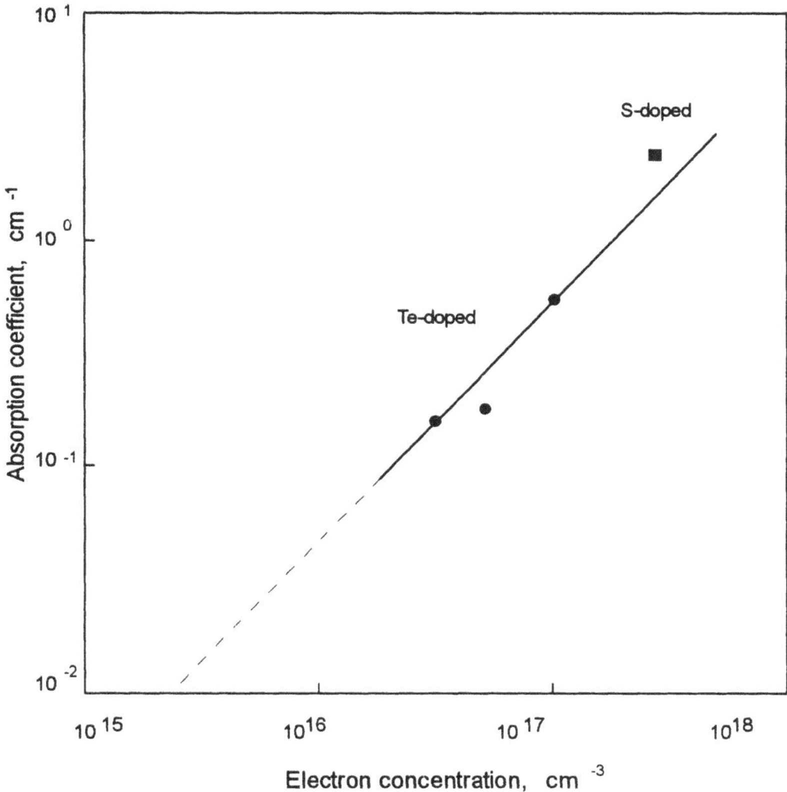

Figure 3.20 Absorption coefficient at 1.1 μm as a function of electron concentration in GaP [5].

We have measured the absorption coefficient of *n*-type GaP at 1.1 μm at room temperature as shown in Figure 3.20 [5].

For the Raman laser experiment, we used *n*-type GaP crystals with carrier concentrations as low as 2×10^{16} cm^{-3}, for which the absorption coefficient is 0.15 cm^{-1}. Therefore, the round-trip absorption loss for a 5-mm long crystal becomes approximately 10% to 14%. For crystals with carrier concentrations exceeding 1×10^{17} cm^{-3}, no lasing has been observed.

3.3 HOMOEPITAXIAL GaP RAMAN LASER

As pointed out in Section 3.1.5, the threshold of the semiconductor Raman laser is very sensitive to crystal quality (e.g., to factors such as free-carrier concentration and deep-level concentration). It is therefore better to use epitaxial GaP than bulk single crystal GaP. However, the maximum attainable thickness of epitaxial layers is limited. First, we tried to grow very high quality and relatively thick GaP epitaxial layers by the liquid phase epitaxial technique called the temperature difference method under controlled vapor pressure (TDM-CVP), which will be described in detail in Chapter 6.

The structure of a homoepitaxial GaP Raman laser with a cubic polarizer is shown in Figure 3.21(a) [5,9]. Epitaxial layers with thicknesses between 350 and 150 μ are grown on nearly (100)-oriented GaP substrates with an electron concentration of $n = 4 \times 10^{17}$ cm^{-3}. The growth temperature is 880°C, growth time is 24 hr, and the applied overpressure of phosphorus is 160 torr, the significance of which will be described in Chapter 6. The surface of the epitaxial layer was optically polished to a thickness of 280 μm because the epitaxial layer had a thickness distribution of approximately 50 μm. The crystal is cut to a length $\ell = 4$ mm and a width of about 3 to 4 mm. Polishing, etching, and antireflection coating processes are the same as described earlier, except that the side wall in the width direction is not treated. The electron concentration of the epitaxial layer is $0.9 \sim 1.0 \times 10^{16}$ cm^{-3}.

Two different cubic polarizers, 1.0 mm and 0.5 mm, have been used. Figure 3.22 shows the lasing characteristics of the epitaxial layer laser. The threshold optical input power is nearly the same as that for the 380- μm thick bulk crystal. The superiority of the epitaxial layer is not clear when the 1.0 mm polarizer is used, but it is seen that the threshold for the epitaxial layer decreases when the 0.5-mm polarizer is used.

Waveguiding in the epitaxial layer occurs because the substrate crystal has a smaller refractive index due to the higher carrier concentration. The advantages of the waveguiding are lost when we use the polarizer, which has a large size relative to the epitaxial layer. Therefore, we have tried a resonator structure without a polarizer. As shown in Figure 3.21(b), high-reflectivity multilayer films are directly coated on a part of the polished end-faces of the epitaxial layer. The high-reflection coating consists of 13 to 15 alternate layers of SiO_2 and TiO_2 with each layer having $\lambda/4$ thickness. The pump beam is introduced at a direction of about 2° with respect to the resonator axis. If the internal absorption loss is low enough, multiple reflections of the pump beam can result in homogeneous pumping inside the resonator. Therefore, the conversion efficiency from the pump power to the Stokes output power can be higher than expected from the transmittance of the resonator mirror.

Figure 3.23 shows the lasing threshold characteristics of the three different Raman lasers. In the case of Raman laser A, for which the reflectivities of the resonator mirrors are higher than 99.5% (made of 15 layers), the threshold pump power density is lower than 1×10^6 W/cm^2, which is the lowest ever attained. Whereas in the case of B and C, with mirror reflectivities of approximately 99% (made of 13 layers), the threshold pump power density is approximately 4×10^6 W/cm^2. It should be noted that second Stokes radiation at the frequency $\omega_{s2} = \omega_0 - 2\omega_{LO}$ is

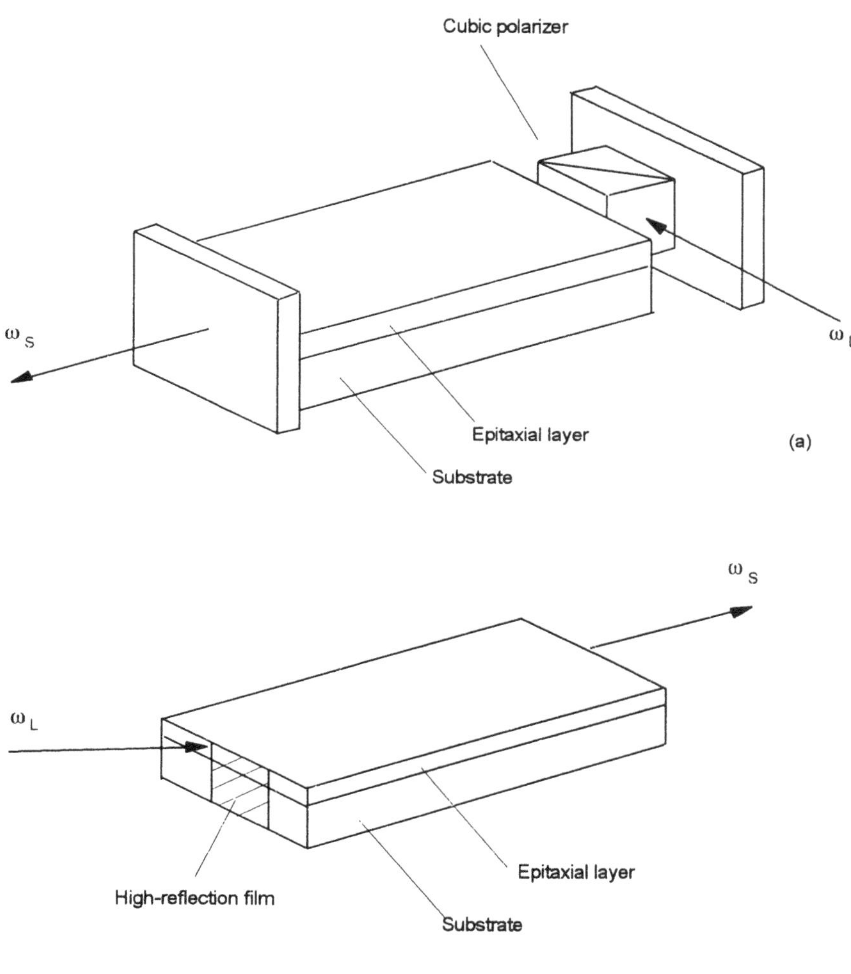

Figure 3.21 Structures of the semiconductor Raman lasers: (a) with external resonator mirrors and a polarizer, (b) with resonator mirrors coated on the crystal surfaces [9].

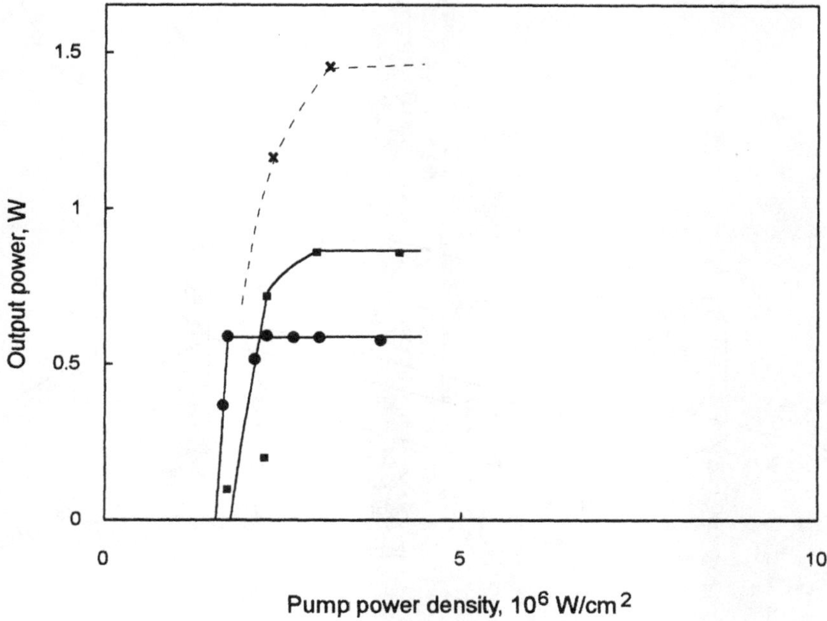

Figure 3.22 Stokes output power as a function of incident optical power density for an epitaxial layer of GaP with thickness 280 μm [5]. (Epitaxial) ■ = 1.0-mm polarizer; ● = 0.5-mm polarizer; (Bulk) = × 1.0-mm polarizer.

also stimulated at a pump power density nearly the same as the threshold for the first Stokes radiation in the case of A and C. The output power of the second Stokes radiation is comparable to, or even exceeds, the output power of the first Stokes radiation.

As shown in Figure 3.24, the reflectivities of the resonator mirrors at the wavelengths of the first and second Stokes lines, 1.112 μm and 1.164 μm, are nearly the same. Therefore, once the intensity of the first Stokes beam exceeds the pump field intensity at threshold, the second Stokes line can also be stimulated in a cascade lasing process. The observed threshold characteristics are, however, not understood by this simple cascade lasing model because the pump threshold for the second Stokes radiation is nearly the same as that for the first Stokes radiation in the case of the Raman laser A and C. This fact means that a coupled interaction system of the pump, the first Stokes, and the second Stokes radiations must be taken into account.

In Section 8.2, we will discuss a general case where two pump radiations with different frequencies, ω_{L1} and ω_{L2}, share the same optical phonons in the stimulated Raman scattering processes. It can be shown that the gain for the coupled system is given by [9]:

$$g = \beta_1 I(\omega_{L1}) + \beta_2 I(\omega_{L2})$$

where β_1 and β_2 are Raman gains for a unit of pump intensity at ω_{L1} and ω_{L2}, and $I(\omega_{L1})$ and $I(\omega_{L2})$ are the pump bean intensities, respectively. The equation implies that the coupled system can exceed the lasing threshold condition even if the intensity $I(\omega_{L2})$ is too weak for lasing, provided the intensity $I(\omega_{L1})$ is high enough. The present case may correspond to a special case of such a coupled system, in which $\omega_{L1} - \omega_{L2}$ coincides the frequency of optical phonons. In reality, the momentum of the optical phonon $\mathbf{q}_2$ causing the second Stokes radiation is not exactly the same as that of the optical phonon $\mathbf{q}_1$ causing the first Stokes radiation because the pump beam direction has a small angle θ_{in} with respect to the resonator axis. In the present experiment, the momentum difference of the two phonons is:

$$|q_1 - q_2| \doteqdot k_1\theta_{in} \doteqdot 2\pi \times 2.4 \times 10^2 \text{ cm}^{-1}$$

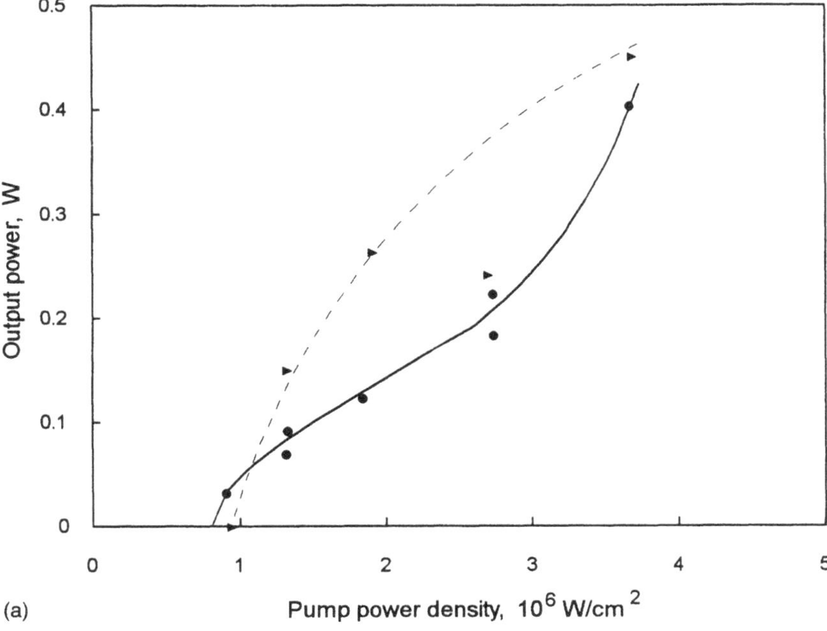

Figure 3.23 Output powers of the Stokes radiations as a function of pump power density [9]. (a) Raman laser A with reflectivities of the coated mirrors R = 99.5% and epitaxial layer thickness t = 80 μm; (b) and (c) Raman lasers B and C, respectively, with R = 99% and t = 90 μm.
● = first Stokes; ▲ = second Stokes; × = third Stokes.

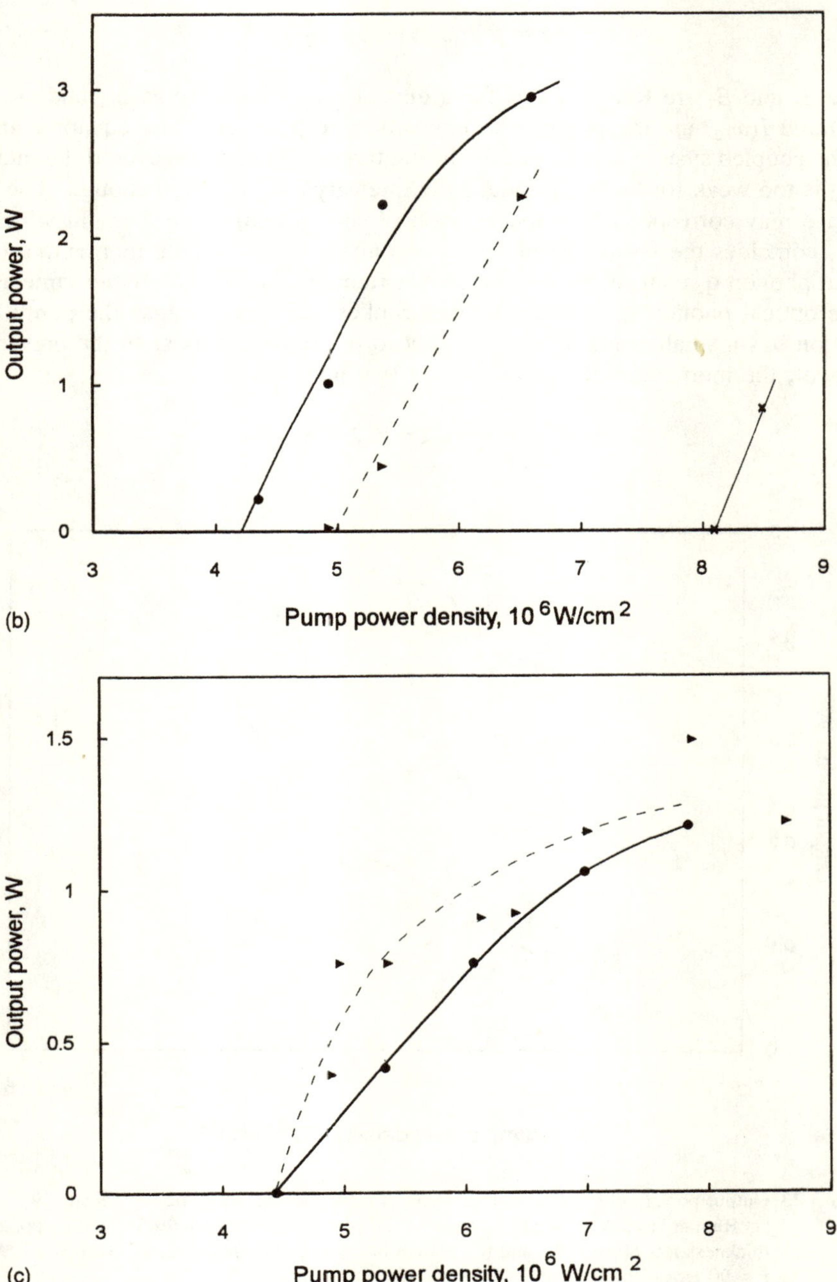

Figure 3.23 (Continued)

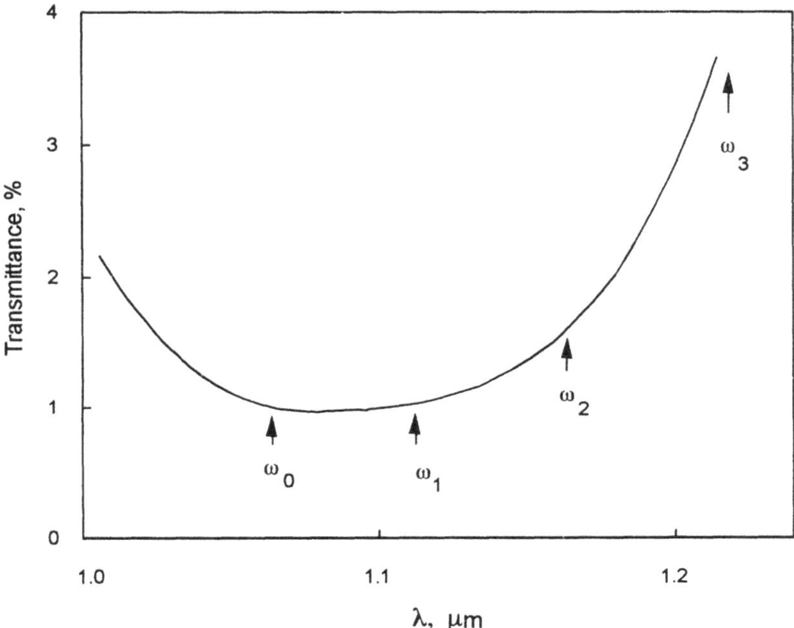

Figure 3.24 Wavelength dependence of the transmission of a coated mirror R = 99% [9].

where k_1 is the momentum of the first Stokes radiation and θ_{in} is the angle of the pump beam direction to the resonator axis. If we assume that this small momentum difference is compensated by the emission of acoustic phonons, the corresponding acoustic phonon frequency is approximately 150 MHz, which is much lower than the natural linewidth of the optical phonon. However, there is the problem of whether a weak optical vibration near the threshold could have such an effect. A more detailed experiment will be necessary to analyze this phenomenon.

To obtain low pump power operation, the pump field should be multiply reflected by the resonator mirrors by making θ_{in} (the angle between the pump beam direction and the resonator axis) as small as possible. Then, only a portion of the pump beam within the width $w_i = 2L \sin \theta_{in}$ can enter the resonator, where L, the length of the Raman laser, is 4 mm in the present experiment. In the case shown in Figure 3.23, θ_{in} was 1.8°. Then, if divergence of the pump beam were to be neglected, the beam would be reflected 4 to 5 times before escaping from the laser, as the resonator width is 1 mm. By decreasing the angle θ_{in}, the incident pump power can be reduced until the internal absorption loss dominates, but only if the resonator has optical confinement in the lateral directions. In the present experiment, however,

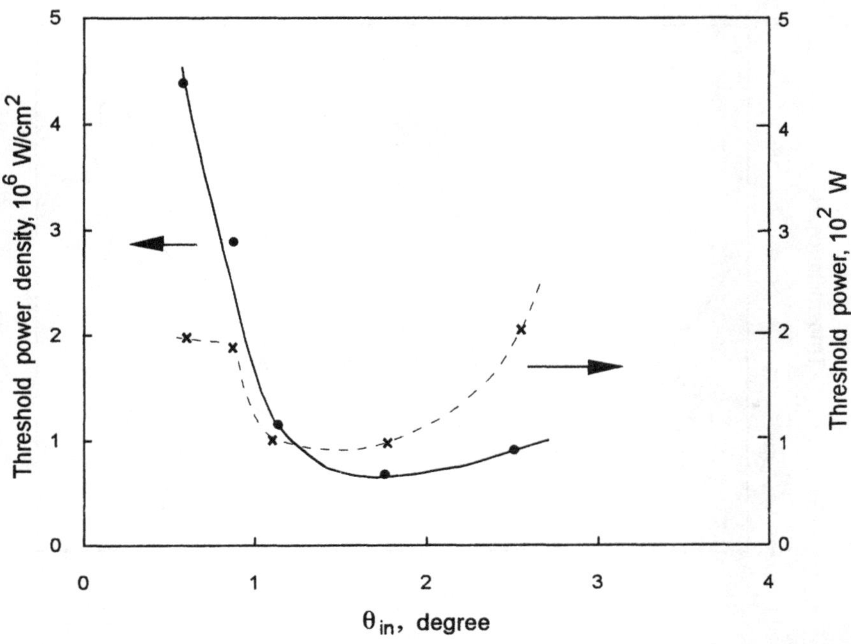

Figure 3.25 Threshold pump power density and the threshold pump power as a function of θ_{in}, the pump beam direction angle inside the crystal relative to the resonator axis [9].

there is no lateral confinement, so homogeneous pumping becomes impossible if θ_{in} is too small. Figure 3.25 shows the θ_{in} dependence of the threshold pump power density, together with the effective incident pump power [9]. The lowest threshold pump power density, 0.8×10^6 W/cm^2, is realized at $\theta_{in} = 1.8°$, at which point the incident pulsed pump power is as low as 100W.

3.4 REGENERATIVE AMPLIFICATION

Light amplification is an important feature of the semiconductor Raman laser. The amplified signal frequency ω_s can be varied by varying the pump light frequency ω_L. In other words, the semiconductor Raman laser is essentially a frequency-selective light amplifier.

The most important application of this feature lies in optical heterodyne demodulation in future wideband optical communication systems [1,9,10]. A wideband modulated optical signal or, alternatively, multiple signal laser beams with slightly different wavelengths arrive at the heterodyne demodulation system, and a narrow band from within the wide modulation band or, alternatively, one particular signal beam from many lines is selected and amplified using a local optical oscillator. Let the frequencies of the signal laser light and the local oscillator be ω_s and ω_L; then the next relation holds:

$$\omega_L = \omega_s + \omega_{ph} \tag{3.10}$$

where ω_{ph} is the longitudinal optical phonon frequency for the Raman laser. For GaP, $\omega_{ph} = 12$ THz. The bandwidth of operation is the gain band of the semiconductor Raman amplifier and corresponds to the linewidth of the spontaneous Raman scattering spectrum. This natural bandwidth is approximately 20 to 30 GHz in the case of semiconductor Raman amplifiers using GaP crystals. We have demonstrated selective amplification, which satisfies (3.10), by using two Raman lasers as shown in Figure 3.26 [9]. The first Raman laser, SR1, is the same as that already described and acts as a signal laser, whereas the second Raman laser, SR2, acts as an amplifier. The latter has a resonator width of 500 μm, and the signal and pump radiations are introduced from different end-faces of the resonator, as illustrated in Figure 3.26. The incidence angle θ_{in} between the signal beam and the amplifier axis was 0.22°, much smaller than that of the pump beam. A YAG laser beam with a pulse width of 40 ns was introduced into both the Raman lasers using a beam splitter.

As shown in Figure 3.27, the output power from the amplifier when the signal radiation is introduced is 2.5 to 5 times higher than the output power level of SR2

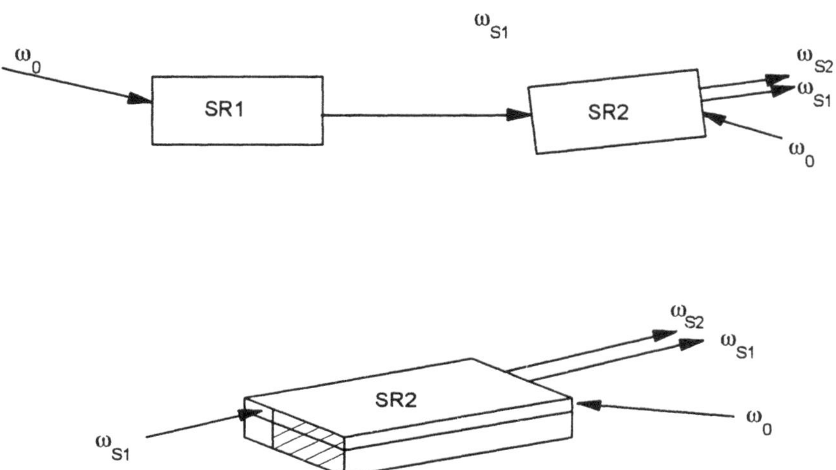

Figure 3.26 Raman laser amplifier for an optical heterodyne demodulation system [9]. SR1 = Raman laser as a signal source; SR2 = Raman amplifier for frequency-band selection.

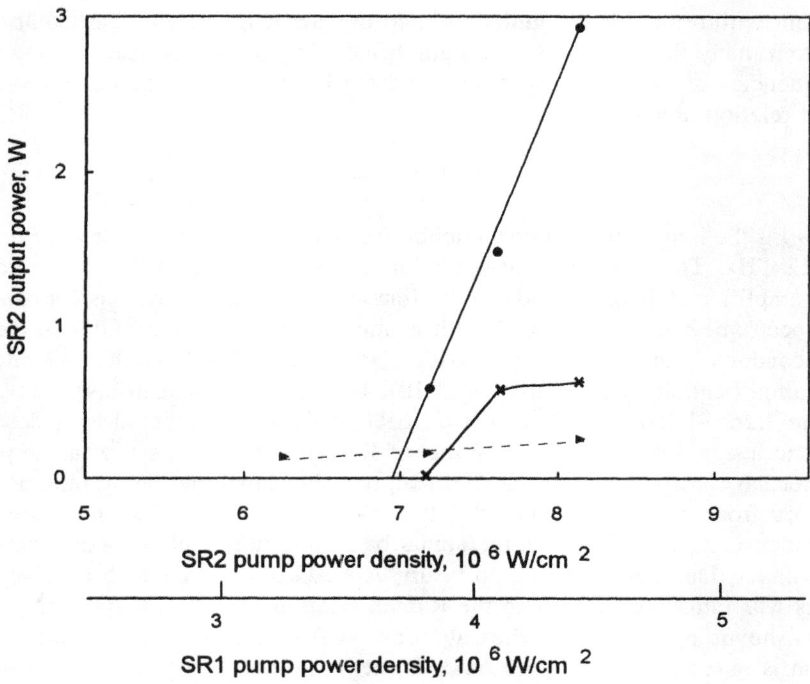

Figure 3.27 Experimental plots of the output power from the Raman laser amplifier SR2 [9]. ● = SR1, SR2 pumped; × = SR2 pumped; ▲ = SR1 pumped.

when there is no signal input. At a pump level just below the threshold of SR2, the amplified output Stokes light is observed only when the signal light is introduced.

The amplifier gain is estimated to be in the range of approximately 5 to 11 dB.

Because θ_{in} of the signal radiation is as small as 0.22°, it would take 15 traverses of the resonator before escaping from the other edge of the resonator mirror if the beam were not diverging. However, a single travelling wave-type amplification will not be the case because the interference between the reflected signal light waves causes an enhancement of amplification, which is called regenerative amplification. Then, the amplification bandwidth should be narrower than the natural bandwidth, approximately 20 to 30 GHz, the center of which is one of the longitudinal modes of SR2. The longitudinal mode separation of a semiconductor Raman laser with 4-mm length is 12 GHz. The output of the signal Raman laser may therefore be a single longitudinal mode or, in the worst case, composed of only two longitudinal modes. Therefore, the frequency difference between the signal radiation and the center frequency of the amplifier should be less than 12 GHz.

It is possible to tune the center frequency of the amplifier or the signal laser by changing the refractive index via the electro-optic effect by a reverse biased pn-junction. More sophisticated amplification experiments will be described in Chapter 7, but the experiment shown here is important because it provides evidence that regenerative light amplification can cause large amplification as a result of internal multiple reflections of the amplified signal light.

Regenerative amplification is a positive feedback effect. Let the amplitude reflectance of the high-reflection resonator mirrors be r_p and r_v, where p and v denote the polarization direction of the light wave parallel or vertical to the plane of incidence, respectively. (For normal incidence, there is no difference between r_p and r_v.) Then, the optical power amplification factor is given by [11]:

$$G_{v,p} = \frac{\exp\{(g - \alpha)l\} \exp(-j2\pi l n_\lambda/\lambda_0)(1 - r_{v,p})^2}{1 - \exp\{2(g - \alpha)l\} \exp\{-j4\pi l n_\lambda/\lambda_0\}(r_{v,p})^4} \tag{3.11}$$

$G_{v,p}$ is maximized for waves satisfying $n_\lambda l/\lambda_0 = 1/2\ m$, where m is an integer. The denominator must be larger than zero so that lasing does not start. It should be noted that the gain-band product remains constant in the regenerative amplification process. Therefore, high gain is obtained at the expense of amplifier bandwidth. The natural bandwidth without any feedback effect is equal to the lattice damping factor Γ. Section 7.3 addresses the amplification factor in more detail.

3.5 MODULATION OF THE SEMICONDUCTOR RAMAN LASER [9]

Changes in the free-carrier concentration cause changes in the optical loss in the Raman laser. Therefore, amplitude modulation can be realized by free-carrier injection from a *pn*-junction. As shown in the inset of Figure 3.28, a *p*-type layer with thickness of approximately 20 μm and hole concentration approximately 1×10^{17} cm^{-3} was grown over the *n*-type epitaxial layer having electron concentration of about 1×10^{16} cm^{-3} and thickness of 70 μm.

The Stokes radiation is confined to the *n*-type epitaxial layer because the substrate and the *p*-type overlayer have slightly lower refractive indices due to higher carrier concentrations. The electrode, which has an area of 3 by 1 mm, was formed using Au-Zn on the surface of the *p*-type layer. The electrode for the *n*-type substrate was formed using Au-Ge, by a standard ohmic contact technique. The forward injection current was applied as a pulse of approximately 1-μs width.

Results are shown in Figure 3.28. In the case of an injection current density $I = 12$ A/cm^2, the effect was too small to be clearly observed. When $I = 25$ A/cm^2, however, a reduction or even quenching of the Stokes output was observed at pump power densities lower than 5×10^6 W/cm^2. This modulation effect could be due to absorption by the injected carriers. If the injected carrier concentration becomes higher than the equilibrium carrier concentration of the active region ($n \simeq 1$

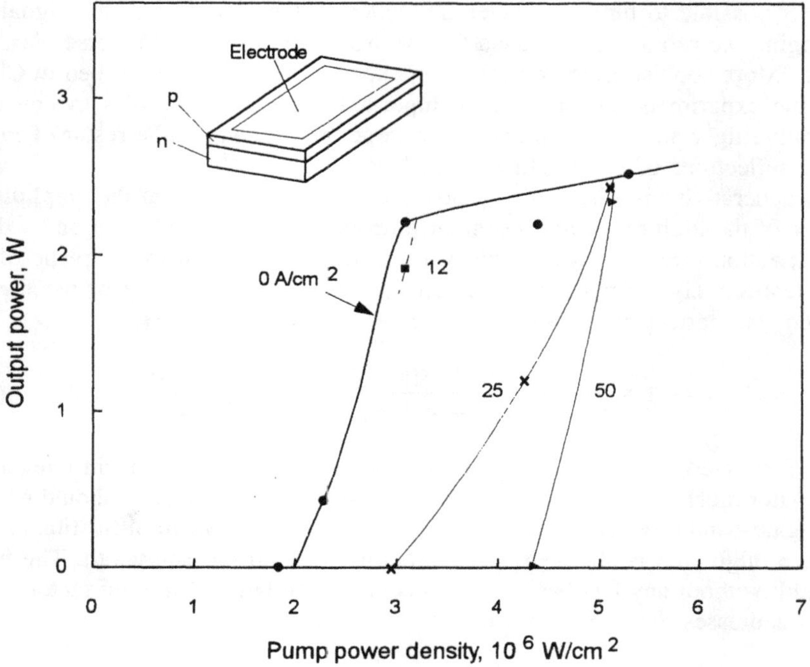

Figure 3.28 Change in the output power of the first Stokes radiation in response to the carrier injection by the *pn*-junction. The current densities are shown in the figure [9].

$\times\ 10^{16}\ cm^{-3}$), the lasing may stop, or the output power may be reduced because the free-carrier absorption is the dominant loss mechanism of the Raman laser.

The current densities required for modulation are much lower than those for direct modulation in semiconductor injection lasers. This is an advantage of the semiconductor Raman laser because the refractive indices of III-V compounds have large temperature coefficients on the order of 1×10^{-4}/C°, which seriously change the injection laser frequency in response to the heat generation by injection. Also, the refractive index changes with the free-carrier concentration. However, use of an external modulator will be much superior to internal modulation in completely eliminating the difficulty of frequency chirping.

3.6 PARAMETRIC GAIN OF THE POLARITON MODE

A polariton mode is the combination of lattice vibration and electromagnetic wave so that there is a possibility of obtaining infrared coherent light from polariton mode oscillation. Actually, far-infrared radiation was observed in a $LiNbO_3$ Raman laser. [12] However, the Raman gain for the polariton mode is relatively small compared to that for the *LO* phonon mode.

For the polariton mode, the frequency varies with the wave momentum, and thus the lattice wave carries its energy along the path of propagation. In such a case, the interaction between the light waves and the lattice wave should be considered as a parametric interaction between the three travelling waves. The parametric Raman interaction for the polariton mode was discussed by Henry and Garrett [13]. Their conclusion is that, while both the nonlinear susceptibility and the infrared absorption by lattice damping become resonantly large at a lattice resonance frequency, ω_0, the parametric gain is kept constant as a function of frequency because both factors compensate each other.

However, it is easily understood that, at a frequency far below the pure *TO* mode frequency, infrared absorption due to free carriers and other processes becomes so strong that the parametric gain is reduced. As a result, the parametric gain shows a maximum at a frequency near the pure *TO* mode frequency [3]. The following case provides an example.

We describe as follows the pump wave, $\mathbf{E}_L$; the Stokes wave, $\mathbf{E}_s$; a lattice wave, $\mathbf{Q}$; and an electromagnetic wave that couples with the lattice wave, $\mathbf{E}_1$:

$$\begin{aligned}
\boldsymbol{E}_L &= (E_L/\sqrt{2}) \exp(j\mathbf{k}_\mathrm{L} \cdot \boldsymbol{r} - j\omega_L t) + c.c., \\
\boldsymbol{E}_s &= (E_s/\sqrt{2}) \exp(j\mathbf{k}_\mathrm{s} \cdot r + \boldsymbol{\gamma} \cdot \boldsymbol{r} - j\omega_s t) + c.c., \\
\boldsymbol{Q} &= (Q/\sqrt{2}) \exp(j\mathbf{q} \cdot \boldsymbol{r} + \boldsymbol{\gamma} \cdot \boldsymbol{r} - j\omega t) + c.c., \\
\boldsymbol{E}_1 &= (\boldsymbol{E}_1/\sqrt{2}) \exp(j\mathbf{q} \cdot \boldsymbol{r} + \boldsymbol{\gamma} \cdot \boldsymbol{r} - j\omega t) + c.c.,
\end{aligned} \tag{3.12}$$

where ω_L, ω_s, and ω are the angular frequencies of the pump, Stokes, and polariton, respectively; $\mathbf{k}_\mathrm{L}$, $\mathbf{k}_\mathrm{s}$, and $\mathbf{q}$ are vectors expressing the wave momentums together with the losses; and $\boldsymbol{\gamma}$ is a vector expressing amplitude gain.

Thus, the nonlinear wave equations and the equation of motion are:

$$\begin{aligned}
\left(\nabla^2 + \frac{\omega_s^2}{c^2}\varepsilon_s\right)E_s &= -\frac{\omega_s^2}{\varepsilon_0 c^2}\alpha_{ij,k}E_L Q^* \\
\left(\nabla^2 + \frac{\omega^2}{c^2}\varepsilon_s\right)E_1 &= -\frac{\omega^2}{\varepsilon_0 c^2}(e_B^* N Q) \\
M(-\omega^2 + \omega_0^2 - j\omega\Gamma)Q &= e_B^* E_1 + \frac{\alpha_{ij,k}}{N}E_L E_s^*
\end{aligned} \tag{3.13}$$

where ω_0 is the frequency of the pure *TO* mode and e_B^*, N, and M are defined in Section 2.1. There is an electro-optic contribution to the Raman scattering that is opposite in sign to the contribution of the lattice vibration. However, this has been neglected because this effect should be small in the present experimental frequency range.

When a strong radiation with frequency ω_L is incident, we solve the secular equation, putting (3.12) into (3.13). On the assumption that $k_L^2, k_s^2 \gg \gamma^2$, the secular equation gives:

$$(2\mathbf{k}_{SO}\cdot\gamma)(2\mathbf{q}_0\cdot\gamma + 2\mathbf{q}_0\beta_I) = \frac{\omega_s^2}{\varepsilon_0 c^2}\frac{|\alpha_{ij,k}|^2}{\rho}\frac{|E_L|^2}{2}\frac{1}{\omega_0^2-\omega^2+j\omega\Gamma} \times \frac{\omega^2}{c^2}\frac{\omega_p^2}{\omega_0^2-\omega^2+j\omega\Gamma} \tag{3.14}$$

where $\mathbf{k}_{SO}$ and $\mathbf{q}_0$ are the real parts of $\mathbf{k}_s$ and $\mathbf{q}$, and β_I is the imaginary part of $\mathbf{q}$ when there is no corresponding gain.

The direction of the vector $\boldsymbol{\gamma}$ can be assumed to be nearly equal to the direction of the propagation of the scattered light when the input angle θ_{in} is small, so the optical power gain $g \approx 2\gamma$.

There are three typical cases in which the gain has different expressions:

1. The gain 2γ is much larger than the loss of the polariton wave; that is, $\gamma \gg \beta_I$. Then, (3.14) gives the expression:

$$g \approx 2\gamma = \frac{1}{\sqrt{\varepsilon_0 k_{SO} q_0 \cos\theta_{IS}}}\frac{\omega_s\omega}{c^2}\frac{|\alpha_{ij,k}|}{\sqrt{\rho}}\frac{|E_L|}{\sqrt{2}}\frac{\omega_p}{(\omega_0^2-\omega^2)} \tag{3.15}$$

where γ, k_{SO}, and q_0 are the magnitude of the vectors $\mathbf{k}_{SO}$, $\mathbf{q}_0$, and $\boldsymbol{\gamma}$, respectively. The gain is proportional to the square root of the incident power density I_L, which corresponds to an ideal parametric case. In the present experiment, in which a high-Q Fabry-Perot resonator is used, the gain cannot exceed 0.1 cm^{-1}, but the infrared absorption exceeds 1 cm^{-1}. Hence, the experimental situation is far from this case.

2. The infrared loss is higher than the gain so that $\beta_I \gg \gamma$, and the lattice absorption is dominant compared to the residual absorption. Then, we have the following expression:

$$g \approx 2\gamma = \frac{\omega_s^2}{\varepsilon_0 k_{SO}\omega\Gamma c^2}\frac{|\alpha_{ij,k}|^2}{\rho}\frac{|E_L|^2}{2} \tag{3.16}$$

The gain is proportional to the input power, but there is no resonance effect when ω becomes close to ω_0 [13].

3. $\beta_1 \gg \gamma$, and the residual absorption is dominant. Then, we have the following expression:

$$g \approx \frac{\omega_s^2 \omega^2}{\varepsilon_0 k_{SO} q_0 c^4 \alpha_l} \frac{\omega_p^2}{(\omega_0^2 - \omega^2)^2} \frac{|\alpha_{ij,k}|^2}{\rho} \frac{|E_L|^2}{2} \tag{3.17}$$

as long as $(\omega_0^2 - \omega^2) \gg \omega\Gamma$.

The Raman gain reduces as $\omega^2/(\omega_0^2 - \omega^2)^2$ when ω departs from the pure transverse phonon frequency ω_0. As will be shown, the Raman gain or the scattering efficiency of the polariton mode in GaP crystals (at least at room temperature) corresponds to this case.

When the incident beam is absent, putting the wave vector of the infrared as $q = q_0 + i\beta_l$, the absorption coefficient in the infrared, $2\beta_l$, is given by:

$$2\beta_l = \alpha_l + \frac{1}{q} \frac{\omega^2}{c^2} \frac{\omega\Gamma\omega_p^2}{(\omega_0^2 - \omega^2)^2 + \omega^2\Gamma^2} \tag{3.18}$$

where α_l is residual absorption due to free carriers, impurities, lattice defects, etc., and is related to the imaginary part of the permitivity ε_l by $\alpha_l = (\omega^2/c^2)(1/q_0)Im\{\varepsilon_l\}$. The second damping term of (3.18) is due to damping of the lattice vibration.

Assuming the residual absorption terms are due to free carriers, Figure 3.29 compares both absorption terms. It clearly shows that if the residual absorption is of the order of 10 cm^{-1} (carrier concentration 2×10^{15} cm^{-3}), it is sufficient to exceed the lattice absorption term throughout the experimental frequency range. At a lower temperature α_l should reduce, but it should be noted that the lattice absorption would also reduce.

Figure 3.30(a) shows the intensity of spontaneous Raman scattering from TO phonons as a function of the incident angle θ_{in} in a crystal with $\varphi = 22°$. As was shown in Figure 3.8, the angular factor f^T is higher than 90% throughout the experimental range of θ_{in}. Therefore, the measured intensities nearly correspond to the dependence of the parametric gain factor. Actually, intensity as a function of θ_{in} can be closely fitted by a function $\omega^2/(\omega_0^2 - \omega^2)^2$, as shown in Figure 3.30(b).

Oscillation of the *TO* polariton mode can be observed when the incident angle is not too small, as shown by the points in Figure 3.7(b). If θ_{in} is too small, the lasing threshold becomes too high; whereas, if θ_{in} is too large, homogeneous excitation along the crystal length becomes difficult. The output power of the Stokes light is still small compared to that of the *LO* phonon mode.

Direct observation of far-infrared power was made by using a zinc-doped Ge detector. The detected power is still very low. One reason is that the optical Stokes

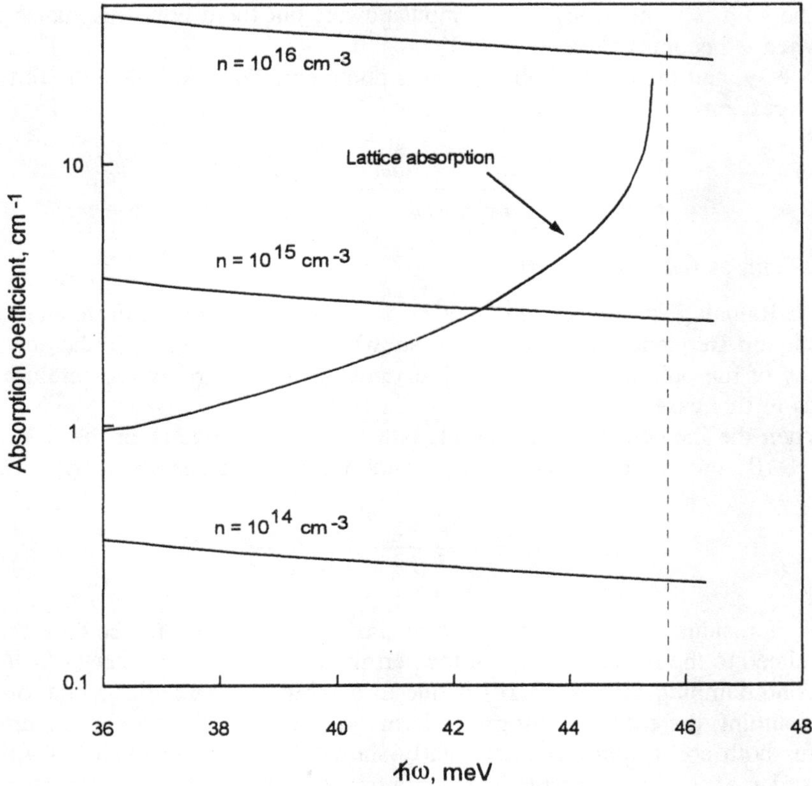

Figure 3.29 Calculated absorption coefficients. — = Free carriers; --- = lattice vibration [3].

output power is not yet high enough. The other reason may be due to transmission loss of far-infrared power from the crystal to the detector, as well as internal absorption loss and reflection at the surface of the crystal.

3.7 GENERATION OF FAR-INFRARED RADIATION

There are some possible methods to generate far-infrared coherent light from lattice vibrations. As discussed in Sections 3.1 and 3.6, the oscillation of the polariton mode will generate far-infrared radiation at the frequency of the polariton mode because the polariton is a mixture of a lattice wave and a far-infrared electromagnetic wave.

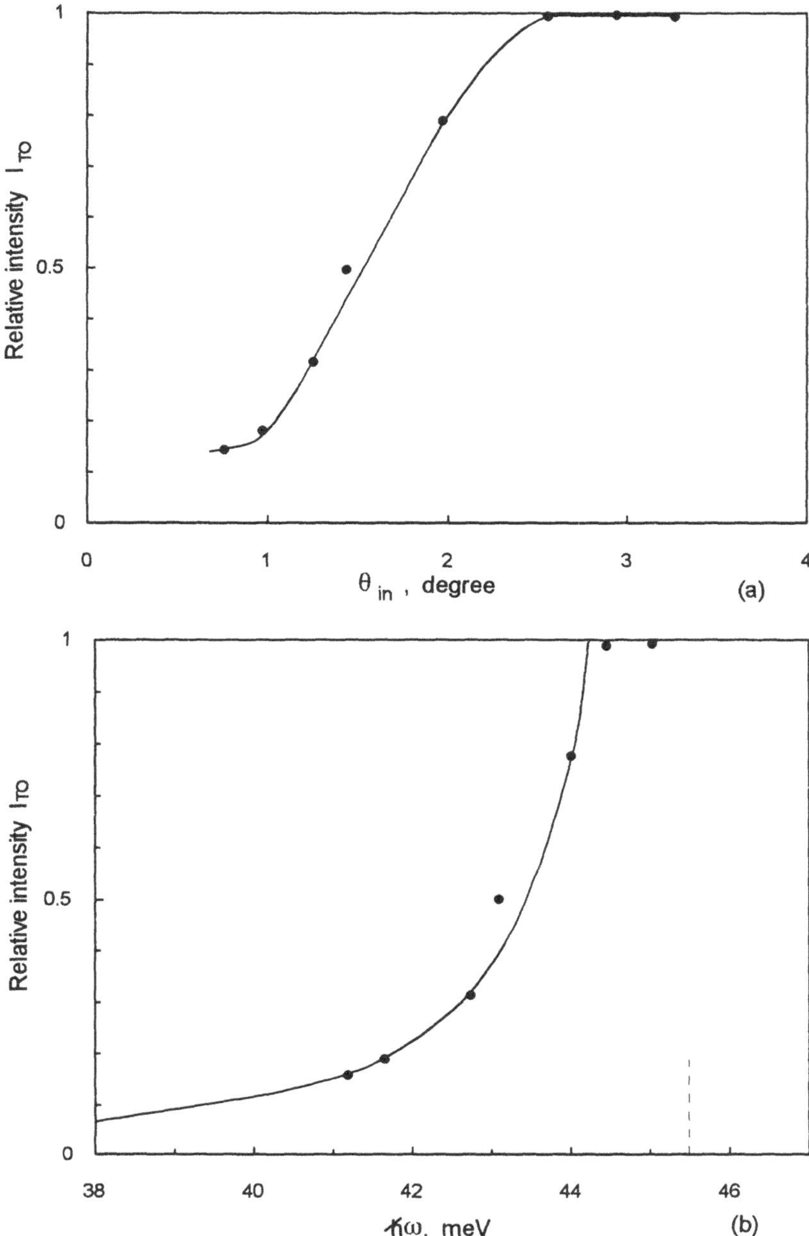

Figure 3.30 Intensity of spontaneous *TO* phonon scattering for $\phi = 22°$ as a function of (a) the incident angle θ_{in}, and (b) the polariton energy [3].

However, its lasing threshold is relatively high, as was discussed. An alternative method is to introduce two laser diode beams with variable frequencies, ω_{L1} and ω_{L2}, to cause difference-frequency mixing. As will be discussed in Chapter 5, a very narrow GaP—$Al_xGa_{1-x}P$ heterostructure waveguide adaptable to laser diode pumping can be fabricated. If the waveguide is aligned along the $\langle 110 \rangle$ crystal direction, the polariton mode wave with frequency $\omega = \omega_{L1} - \omega_{L2}$ can nearly satisfy the phase matching condition for colinear interaction $\theta_{in} = 0$, in a certain frequency range, provided $\omega < \omega_0$. Referring to Figure 3.7, it is found that, for example, when $\lambda_{L1} = 1.064$ μm, approximate phase matching can be obtained in the range 20 meV $\leq \hbar\omega \leq 30$ meV; that is, in the far-infrared wavelength range 40 μm $\leq \lambda \leq 60$ μm, and that the phase matching frequency for collinear interaction decreases as both of the pump light frequencies ω_{L1} and ω_{L2} are decreased.

Compared to the high threshold of the polariton mode, the threshold of the *LO* phonon mode is much lower. It is, therefore, more practical to develop methods to generate far-infrared radiation from the *LO* phonon mode oscillation [4]. We have investigated a method to cause difference-frequency mixing in the resonator. As shown in Figure 3.32, a semiconductor crystal other than GaP that has a *TO* phonon frequency ω_{TO} a little higher than the *LO* phonon frequency of GaP ($\hbar\omega_{LO}$(GaP) = 49.9 meV) is placed, together with GaP, in a resonator with an alignment that enables its *TO* phonon mode to be excited. By this combination, far-infrared radiation (24.8 μm) can be produced from the mixer crystal with low-threshold input optical power because of the *LO* phonon mode oscillation of GaP.

One of the most suitable semiconductors is AlP, which has a *TO* phonon energy of 54.4 meV. In this section, however, we only demonstrate the effectiveness of this method by using GaAs. The optical phonon energies of GaAs are so small (ω_{TO} = 33 meV, ω_{LO} = 36 meV) that phase matching, such as illustrated in Figure 3.31, cannot be obtained. Nevertheless, we can expect to detect far-infrared radiation due to the advantage given by the small absorption coefficient of GaAs at 49.9 meV ($\alpha \approx 0.2$ cm^{-1}).

The experimental arrangement is shown in Figure 3.32. A GaAs crystal with a thickness of 260 μm is placed in an optical resonator at Brewster's angle θ_B = 74°. The normal to the surfaces of the GaAs plate is inclined 16° from a (110) plane so that the Stokes radiation from GaP propagates in a $\langle 110 \rangle$ direction in the GaAs plate.

The GaP crystal is placed in the resonator with the $\langle 001 \rangle$ direction parallel to the resonator axis and the $\langle 110 \rangle$ direction parallel to the polarization direction of the pump radiation $\boldsymbol{e}_L$. Thus, the polarization direction of the Stokes radiation $\boldsymbol{e}_s$ becomes parallel to the $\langle 110 \rangle$ direction, so both the pump and Stokes radiation pass through the GaAs without reflection. In reality, the thickness of GaAs is limited by absorption at the 1-μm region ($\alpha \approx 1.5$ cm^{-1} at 1.1 μm).

We have used resonator mirrors with the highest attainable reflectivity, $R_1 = R_2 = 99.5\%$, because a higher internal electric field is more effective for mixing.

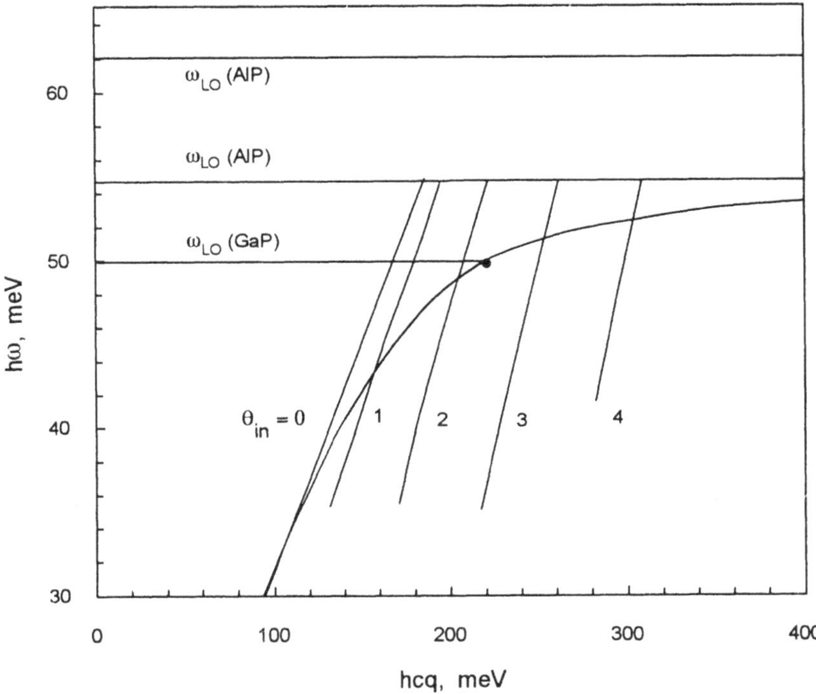

Figure 3.31 Optical phonon dispersion curve of AlP. The *LO* phonon mode oscillation of GaP provides a difference-frequency mixing at a point indicated by solid circle [4].

The far-infrared radiation output power is guided by a copper pipe with a diameter of 10 mm, passed through a Ge low-pass filter and a 19- to 26-μm bandpass filter, and detected by a thermocouple.

Figure 3.33 shows the relation of the total far-infrared radiation power guided by the copper pipe to the pump power.

The maximum far-infrared radiation power is 3W. The low efficiency of the power conversion to far-infrared radiation ($\eta \approx 3 \times 10^{-6}$) is inevitable because the phase matching in GaAs is poor. The wave vector mismatching Δq is given by:

$$\Delta q = \left\{ n_{FIR} - n_i \left(1 - \frac{\lambda_i}{n_i} \frac{dn_i}{d\lambda} \right) \right\} \frac{\omega_{FIR}}{c} \tag{3.19}$$

where n_{FIR} and n_i are reflective indexes at the far-infrared radiation and pump wavelengths. The coherence length L_{coh}, which is given by $L_{coh}\Delta q = \pi$, is as short as 16

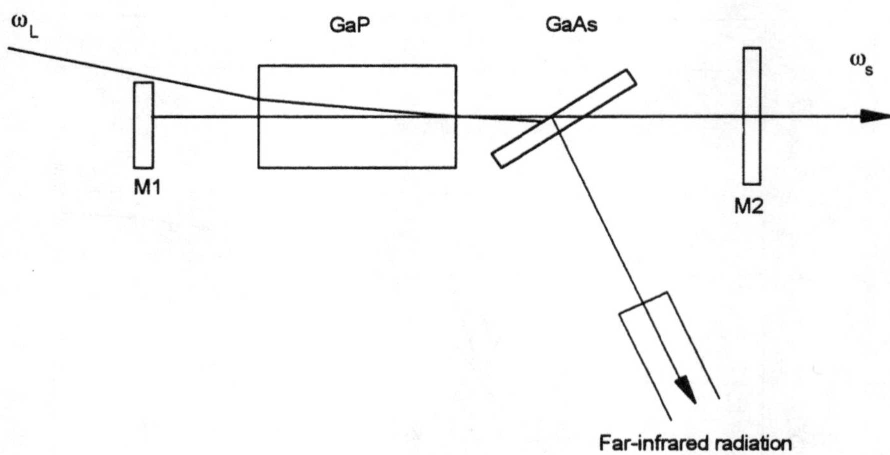

Figure 3.32 Experimental setup for the far-infrared generation employing a GaP oscillator and a GaAs mixer [4].

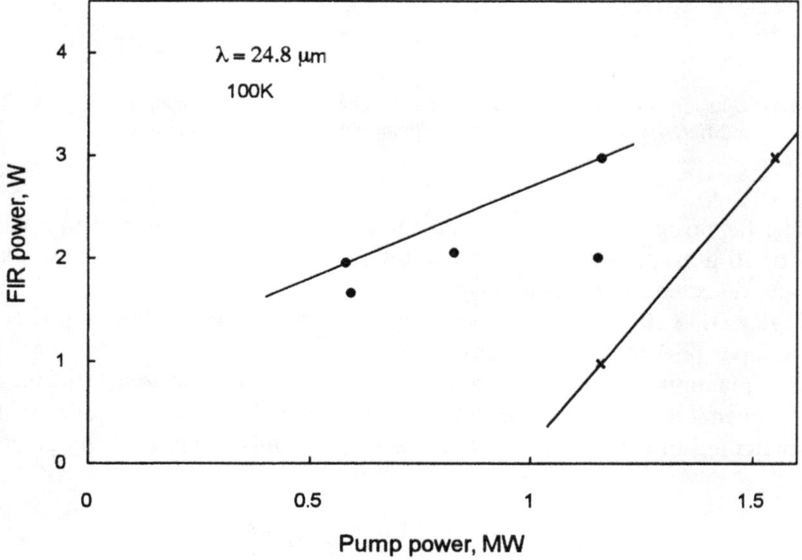

Figure 3.33 Output power of the far-infrared radiation (24.8 μm) as a function of the incident power at 100 K [4].

μm. The detected far-infrared radiation power should, therefore, be some average that has not been cancelled out along the thickness direction.

As suggested in Figure 3.31, we can expect perfect phase matching in AlP, where a high-conversion efficiency on the order of 1% is expected. The good lattice matching between GaP and AlP would be advantageous in an integrated structure.

REFERENCES

[1] J. Nishizawa and K. Suto, "Semiconductor Raman Laser," *J. Appl. Phys.*, Vol. 51, 1980, pp. 2429–2431.

[2] J. Nishizawa and K. Suto, "Optical Wave Demodulator," Japanese Patent 1605283, 1981.

[3] J. Nishizawa and K. Suto, "Semiconductor Raman and Brillouin Lasers for Far-Infrared Generation," *Infrared and Millimeter Waves*, Vol. 7, ed. K. J. Button, New York: Academic Press, 1983, pp. 301–320.

[4] K. Suto and J. Nishizawa, "Low-Threshold Semiconductor Raman Laser," *IEEE J. Quantum Electron*, Vol. QE-19, 1983, pp. 1251–1254.

[5] K. Suto and J. Nishizawa, "Semiconductor Raman Laser," *IEE Proc.*, Vol. 132, Pt. J, 1985, pp. 81–84.

[6] M. Born and E. Wolf, *Principles of Optics*, New York: Pergamon, 1964, p. 430.

[7] D.A. Kleinman and W.G. Spitzer, "Infrared Lattice Absorption of GaP," *Phys. Rev.*, Vol. 118, 1960, pp. 110–117.

[8] W.G. Spitzer, M. Gershenzon, C.J. Frasch and D.F. Gibbs, "Optical Absorption in *n*-type Gallium Phosphide," *J. Phys. Chem. Solids*, Vol. 11, 1959, p. 339.

[9] K. Suto and J. Nishizawa, "Characteristics of the Epitaxial Semiconductor Raman Laser," *IEE Proc.*, Vol. 133, Pt. J, 1986, pp. 259–263.

[10] K. Suto, S. Ogasawara, T. Kimura and J. Nishizawa, "Semiconductor Raman Laser as a Tool for Wideband Optical Communication," *IEE Proc.*, Vol. 137, Pt. J, pp. 43–48.

[11] J. Nishizawa, *Optoelectronics* (in Japanese), Tokyo: Kyoritzu Press, 1977.

[12] J.M. Yarborough, S.S. Sussman, H.E. Purhoff, R.H. Pantell, and B.C. Johnson, "Efficient Tunable Optical Emission from $LiNbO_3$ without a Resonator," *Appl. Phys. Lett.*, Vol. 15, 1969, pp. 102–105.

[13] C.H. Henry and C.G.B. Garrett, "Theory of Parametric Gain near a Lattice Resonance," *Phys. Rev.*, Vol. 171, 1968, pp. 1058–1064.

Chapter 4
Semiconductor Brillouin Laser

4.1 PRINCIPLE OF THE STIMULATED BRILLOUIN SCATTERING

The lattice vibration with acoustic mode (i.e., the sound wave) interacts with the light wave. This is called the Brillouin scattering. Similar to the discussion for the stimulated Raman scattering, the stimulated Brillouin scattering occurs as an amplifying process for the Stokes-shifted light when a strong pump light is incident. Together with the frequency relation

$$\omega_L = \omega_s + \omega_{ac} \tag{4.1}$$

we must take into account the momentum conservation relation:

$$\mathbf{k}_L = \mathbf{k}_s + \mathbf{q} \tag{4.2}$$

where ω_{ac} and $\mathbf{q}$ are the frequency and the wave momentum of the acoustic wave, respectively.

As long as the wavelength of an acoustic wave is much larger than the lattice spacing, the velocity of an acoustic wave, v_{ac}, is independent of frequency:

$$\omega_{ac} = v_{ac} q \tag{4.3}$$

Then, we have the following equation from the momentum relation (4.2) and the assumption $|k_L| \simeq |k_s|$ $(\omega_L \simeq \omega_s)$,

$$\omega_{ac} = \frac{2v_{ac}}{c} n_L \omega_L \sin\frac{\theta}{2} \tag{4.4}$$

where θ is an angle between the propagation directions of the pump light and the scattered light, as illustrated in Figure 4.1.

The frequency ω_{ac} for the Brillouin scattering becomes a maximum at $\theta = 180°$; that is, for the backward scattering. This is the most interesting case for the Brillouin laser because the collinear interaction is obtained at $\theta = 180°$. The Brillouin shift for the backward scattering, $\omega_{max} = 2v_{ac}/c\, n_L\omega_L$, is in proportion to the light frequency.

As shown in Figure 4.2, at a light wavelength of 1 μm, $\omega_{\max}$ for various solid materials is in the range of 10 to 100 GHz [1]. The importance of this range of the frequency shift was not so well recognized when the linewidth of a pump laser light was not narrow enough. However, if we have Brillouin shifted light sources with high stability and high coherency, they will serve optical communication in the future as a local oscillator for an optical heterodyne demodulator or as a closely frequency shifted signal source.

For describing the amplifying process in the stimulated Brillouin scattering, we start from the nonlinear free energy density function, F^{NL}, similar to (2.28) for the Raman scattering:

$$F^{NL} = \left\{\frac{\varepsilon^2}{\varepsilon_0} P_{ijkl}E_i(\omega_L)E_j^*(\omega_s)S_{kl}^*(\omega_{ac}) + c.c.\right\} \tag{4.5}$$

where S_{kl} is the strain tensor, and P_{ijkl} is the photoelastic constant. The factor $\varepsilon^2/\varepsilon_0$ appears because the photoelastic constant is defined in terms of **D** instead of **E**.

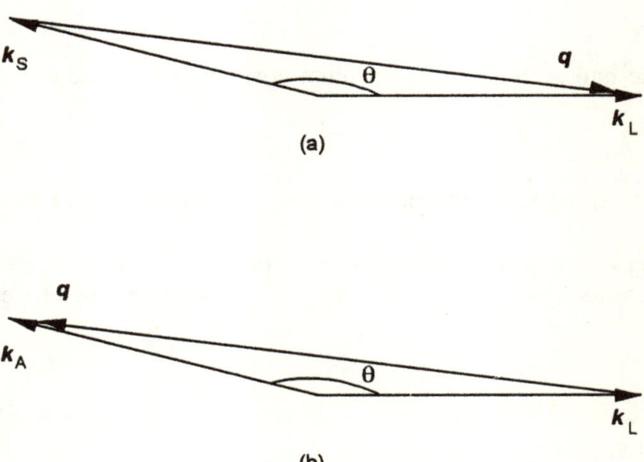

Figure 4.1 Momentum relations in Brillouin scattering: (a) Stokes scattering, (b) anti-Stokes scattering.

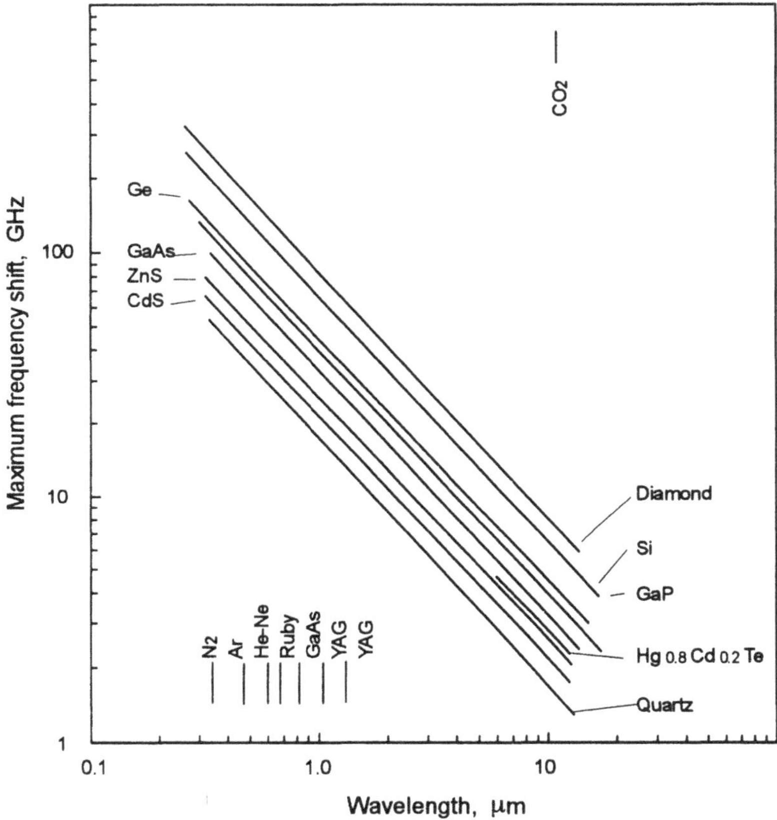

Figure 4.2 Brillouin shift for the backward scattering [1].

The change in the dielectric constant at an optical frequency caused by the strain S_{kl} is given by:

$$\Delta\left(\frac{1}{\varepsilon_{ij}}\right) = -\frac{\Delta\varepsilon_{ij}}{\varepsilon^2} = \frac{1}{\varepsilon_0} P_{ijkl} S_{kl} \tag{4.6}$$

Let us consider the longitudinal acoustic wave propagating in the z direction in a cubic crystal. Using the engineering notations for the suffixes ij and kl of the tensor components, $11 \to 1$, $22 \to 2$, $33 \to 3$, $23 \to 4$, $13 \to 5$, and $12 \to 6$, we have:

$$\Delta\varepsilon = -\frac{\varepsilon^2}{\varepsilon_0} p_{23}(S_3 e^{jqz - j\omega_{ac}t} + c.c.) \tag{4.7}$$

From the Maxwell equation (2.25), the propagation equations for the pump and Stokes light waves are given by:

$$\left(\nabla^2 + \frac{\omega_L^2}{c^2} n_L^2\right) E(\omega_L) = \frac{\omega_L^2}{\varepsilon_0 c^2} \frac{\varepsilon^2}{\varepsilon_0} P_{23} E(\omega_s) S_3(\omega_{ac}) \tag{4.8}$$

$$\left(\nabla^2 + \frac{\omega_s^2}{c^2} n_s^2\right) E(\omega_s) = \frac{\omega_s^2}{\varepsilon_0 c^2} \frac{\varepsilon^2}{\varepsilon_0} P_{23} E(\omega_L) S_3^*(\omega_{ac}) \tag{4.9}$$

The propagation of the longitudinal acoustic wave can be described as:

$$\rho \ddot{S}_3 - C_{33} \nabla^2 S + \Gamma \rho \dot{S} = \frac{\partial^2}{\partial z^2} T^{NL} \tag{4.10}$$

where Γ is the damping constant, ρ is the mass density, C_{33} is the elastic constant of the longitudinal wave, and T_3^{NL} is the nonlinear stress caused by the light fields, which is given by $T^{NL} = -\partial F^{NL}/\partial S_3$. Then we have:

$$\rho \ddot{S}_3 - C_{33} \nabla^2 S_3 + \Gamma \rho S_3 = q^2 \frac{\varepsilon^2}{\varepsilon_0} P_{23} E(\omega_L) E^*(\omega_s) \tag{4.11}$$

Assuming the following forms, $E(\omega_L) = E_L(z)\, e^{jkLz - j\omega Lt}$ + c.c., $E(\omega_s) = E_s(z)$ $e^{jk_s z - j\omega_s t}$ + c.c., and $S_3(\omega_{ac}) = S_3(z)\, e^{jqz - j\omega_{ac}t}$ + c.c., the equations (4.8), (4.9), and (4.10) can be written as:

$$\frac{\partial E_L}{\partial z} = -\frac{1}{2jk_L} \frac{\omega_L^2}{\varepsilon_0 c^2} \frac{\varepsilon^2}{\varepsilon_0} P_{23} E_s S_3 \tag{4.12}$$

$$\frac{\partial E_s}{\partial z} = \frac{1}{2jk_s} \frac{\omega_s^2}{\varepsilon_0 c^2} \frac{\varepsilon^2}{\varepsilon_0} P_{23} E_L S_3^* \tag{4.13}$$

$$2jqC_{33} \frac{\partial S_3}{\partial z} + \Gamma \rho j \omega_{ac} S_3 = q^2 \frac{\varepsilon^2}{\varepsilon_0} P_{23} E_L E_s^* \tag{4.14}$$

When the Stokes field is so weak that the pump field depletion does not occur, E_L can be assumed to be constant.

For the acoustic wave propagation, we consider two different cases.

The first is when the damping term $\Gamma\rho j\omega_{ac}S_3$ is much larger than the amplification term $2jqC_{33}\ \partial S_3/\partial z$. Then, S_3 is given by:

$$S_3 = \frac{1}{\Gamma\rho j\omega_a} q^2 \frac{\varepsilon^2}{\varepsilon_0} P_{23}E_L E_s^* \tag{4.15}$$

In this case, (4.13) becomes:

$$\frac{\partial E_s}{\partial z} = \left\{\frac{k_s q}{2\varepsilon}\left(\frac{\varepsilon^2}{\varepsilon_0} P_{23}\right)^2 |E_L|^2 \frac{v_{ac}}{\Gamma C_{33}}\right\} E_s = \left\{\frac{k_s q}{2\varepsilon}\left(\frac{\varepsilon^2}{\varepsilon_0} P_{23}\right)^2 |E_L|^2 \frac{L_{ac}}{C_{33}}\right\} E_s \tag{4.16}$$

where $L_{ac} = v_{ac}/\Gamma$ is the attenuation length of the acoustic power. We have used the relations $C_{33} = \rho v_{ac}^2$ and $\varepsilon = n^2\varepsilon_0$.

Therefore, the power gain for unit length for the Stokes light is given by:

$$g = \frac{1}{\varepsilon}\left(\frac{\varepsilon^2}{\varepsilon_0} P_{23}\right)^2 |E_L|^2 \frac{k_s q v_{ac}}{C_{33}\Gamma} = \frac{1}{\varepsilon}\left(\frac{\varepsilon^2}{\varepsilon_0} P_{23}\right)^2 |E_L|^2 \frac{k_s q L_{ac}}{C_{33}} \tag{4.17}$$

$|E_L|^2$ can be expressed in terms of the light power density P_L by the relation $P_L = 2\varepsilon\ |E_L|^2 c/n_L$.

Conversely, if the damping term of the acoustic wave is much smaller than the amplifying term, we have instead of (4.15):

$$\frac{\partial S}{\partial z} = \frac{q}{2jC_{33}}\left(\frac{\varepsilon^2}{\varepsilon_0} P_{23}\right) E_L E_s^* \tag{4.18}$$

Then, from (4.13):

$$\frac{\partial^2 E_s}{\partial z^2} = \frac{k_s q}{4\varepsilon_0 n_s^2}\left(\frac{\varepsilon^2}{\varepsilon_0} P_{23}\right)^2 |E_L|^2 \frac{1}{C_{33}} E_s \tag{4.19}$$

Therefore, $E_s(z) \propto e^{g/2z}$ dependence is obtained, and the gain for unit length g is given by:

$$g = \frac{1}{\sqrt{\varepsilon}}\left(\frac{\varepsilon^2}{\varepsilon_0} P_{23}\right)|E_L| \sqrt{\frac{k_s q}{C_{33}}} \tag{4.20}$$

In contrast to (4.17), it is found that g is in proportion to the square root of the pump power. The similar dependence on the optical power has also been encountered in the polariton propagation discussed in Chapter 3.6.

The square root dependence should be observed when the pump light power is very high. The square root dependence should also be observed if the attenuation of the acoustic power is effectively reduced by the presence of another acoustic wave amplification mechanism, such as acoustic wave amplification by drifting conduction electrons or acoustic wave generation accompanying Gunn oscillation [2].

In the case of the transverse acoustic wave in a cubic crystal, P_{23} and C_{33} in the above equations should be replaced by P_{44} and C_{44}, respectively.

From (4.17), g is found to be in proportion to a parameter $M_{mn} = n_s^7 P_{mn}^2 / C_{m'n'}$ and the damping constant Γ. Parameters P_{mn}, $C_{m'n'}$, M_{mn}, and Γ for a few typical solid materials are listed in Table 4.1.

Photoelastic constants P_{mn}'s depend on the wavelength of the pump light. For example, Figure 4.3 shows P_{44} for CdS as a function of λ [3]. It should be noted that P_{44} becomes 0 and changes the sign at a photon energy below the bandgap energy. This is because the term corresponding to the resonance enhancement at the bandgap energy has an inverse sign to the nonresonant part.

The damping rate of the acoustic wave is thought to increase with increasing frequency approximately as $\Gamma \propto \omega_{ac}^2$ at a microwave frequency region.

Although the values of M_{mn}'s in the quartz and fused silica are not high compared to those in semiconductors, the damping constant is as small as approximately

Table 4.1
Brillouin Scattering Parameters

Compound	P_{mn} *(λ μm)*	C_{mn} 10^{11}*dyn* cm^{-2}	n_λ	M_{mn} 10^{-11}*dyn*$^{-1}$*cm*^{2}	$\Gamma/2\pi$
GaP	$P_{12} = -0.082$ $P_{44} = -0.074$ (0.63)	$C_{11} = 14.12$ $C_{44} = 7.047$	3.31	2.1 3.4	
GaAs	$P_{12} = 0/066$ $P_{44} = -0.05$ (1.15)	$C_{11} = 11.81$ $C_{44} = 5.94$	3.37	8.2 4.3	
CdS	$P_{12} = 0.066$ $P_{44} = -0.05$ (0.63)	$C_{11} = 9.38$ $C_{44} = 1.58$	2.44	0.24 0.81	0.7 GHz
Fused Quartz	$P_{12} = 0.27$ $P_{44} = -0.075$ (0.63)	$C_{11} = 8.67$ $C_{44} = 5.82$	1.46	0.12 0.014	15–100 MHz

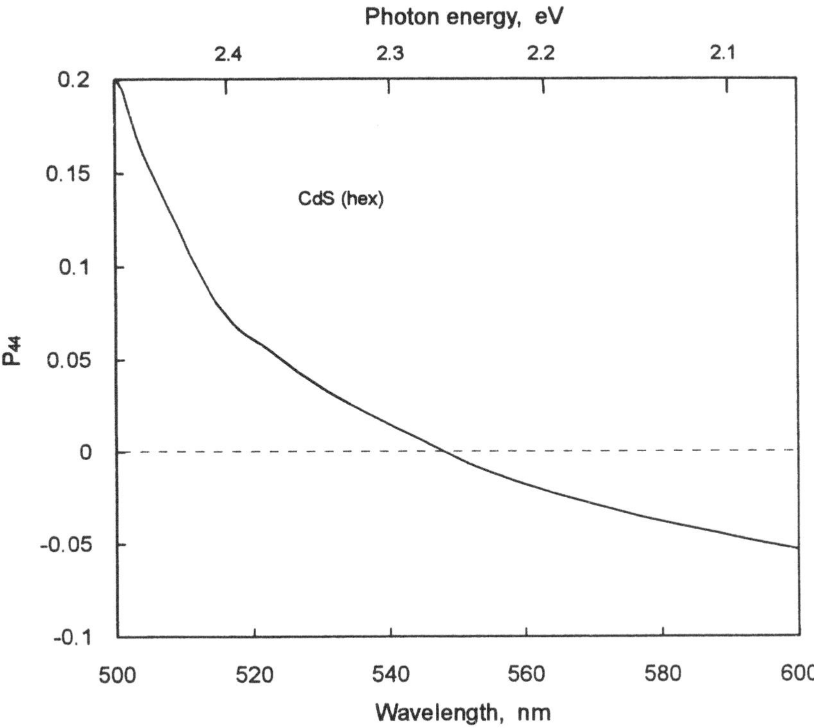

Figure 4.3 Dispersion of the P_{44} photoelastic coefficient in CdS at 77 K, as a function of wavelength [3].

100 to 15 MHz at $\omega_{ac} \simeq 30$ GHz, so the gain of the stimulated Brillouin scattering in the quartz is comparable or higher than those in semiconductors.

4.2 EXPERIMENTS ON THE SEMICONDUCTOR BRILLOUIN LASER

As a basis of the Brillouin laser, the backward scattering of the light wave by the acoustic wave is a very important topic to investigate . When a coherent acoustic wave is excited in a crystal, the Brillouin scattering is the same as the acousto-optic diffraction of the light wave. There are a number of experiments on the acousto-optic diffraction at a relatively low frequency with $\theta \ll 180°$ [4]. However, there are few experiments for the backward acousto-optic diffraction, or the backward Brillouin scattering, because the efficiency of generation of the ultrasonic wave at a frequency higher than 10 GHz is quite low.

The following is our experiment on a CdS crystal [5]. We used a He-Ne laser with a wavelength of 6,328A as an incident light source. Although the power level was low, it was an advantage that the linewidth of the laser was much smaller than the Brillouin linewidth. The Brillouin shift in CdS is 34.8 GHz at this wavelength. To our knowledge, this is the highest frequency acoustic wave yet employed in an acousto-optic diffraction experiment. The experiment on the collinear interaction of the light wave and the microwave sound was performed by the system illustrated in Figure 4.4.

Longitudinal elastic waves are generated at the end surface of a rectangular rod of a CdS crystal measuring 1.0 by 1.0 by 5.0 mm, with the *c* axis along its length, placed in a high electric field portion of a coaxial re-entrant cavity. The cavity is excited by continuous microwave power from a 600-mW klystron. The He-Ne laser emits approximately 5 mW at 6,328 Å, which is incident upon the sample through the small bore at the end of the microwave cavity. The CdS sample, which is the sulfur-compensated nominally pure ($\rho_0 > 10^8\ \Omega$ cm) crystal, is oriented for backward diffraction from the longitudinal acoustic wave along the *c* axis. In this geometry, the backward diffracted light is unchanged in polarization and results from the $P_{13} = P_{23}$ photoelastic tensor component.

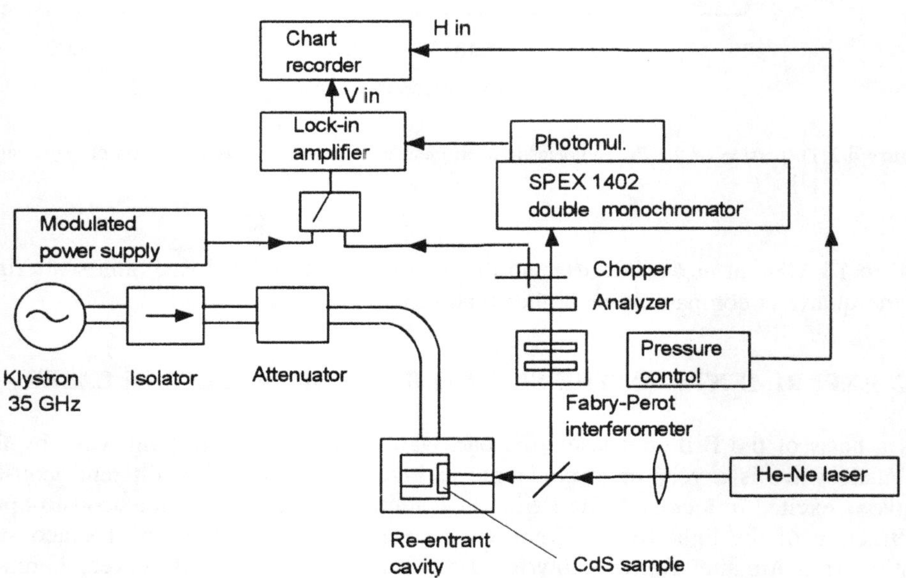

Figure 4.4 Schematic diagram of the experimental arrangement for acoustic-optic diffraction in the backward direction in CdS [5].

The Stokes component is resolved by a plane Fabry-Perot interferometer with a free spectral range of 25 GHz. The strong unshifted light scattered from the surfaces of the crystal necessitated a SPEX 1402 double monochromator as a prefilter. The light diffracted is chopped at 176 Hz before being focused into a double monochromator so that the photomultiplier output can be applied to a lock-in amplifier with output that provides the *Y* input for an *X-Y* recorder as seen in the trace of Figure 4.5(a). The diffracted light polarization is examined by the analyzer and is found to be parallel to the incident light polarization.

To improve the overall contrast, the reference signal for the lock-in amplifier is supplied by a klystron whose amplitude is square wave modulated at 2.1 KHz with the chopper off. The Brillouin spectra obtained at 295K by varying the microwave frequency are shown in Figure 4.5(b). The magnitudes of Brillouin shifts in Figure 4.5(b) are in agreement with the input microwave frequencies.

The diffracted light intensity corrected for the frequency dependence of the electric field in the cavity is plotted in Figure 4.6. The spectrum has a peak at 34.8 GHz, which is in agreement with the calculated value of the maximum frequency shift and also with the thermal phonon frequency experimentally obtained. The full width at half-maximum of the spectrum is about 1.4 GHz, but the actual linewidth may be somewhat narrower than this value because the spectra in Figure 4.6 may include the nonresonant contribution.

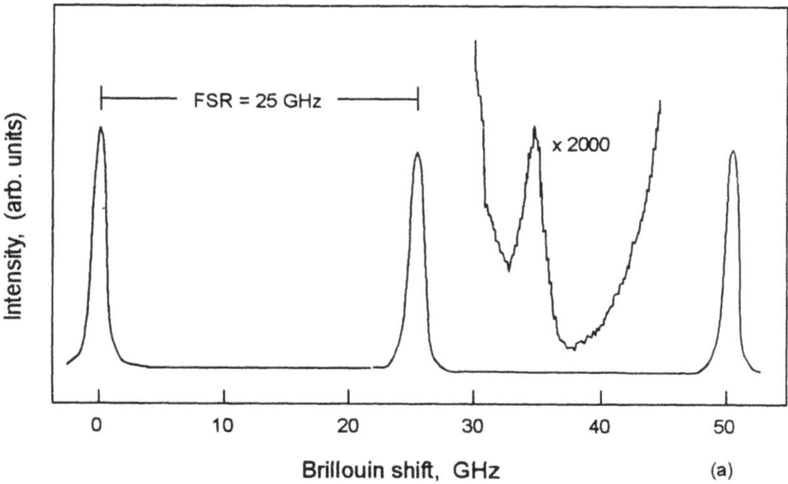

Figure 4.5 Brillouin spectra: (a) Brillouin backscattering spectrum in CdS at room temperature detected with a lock-in amplifier whose reference signal is supplied by the rotating chopper; (b) variation of Brillouin spectrum for different values of input microwave frequencies at room temperature. The reference signal for the lock-in amplifier was obtained from the modulated krystron power supply with the chopper off [5].

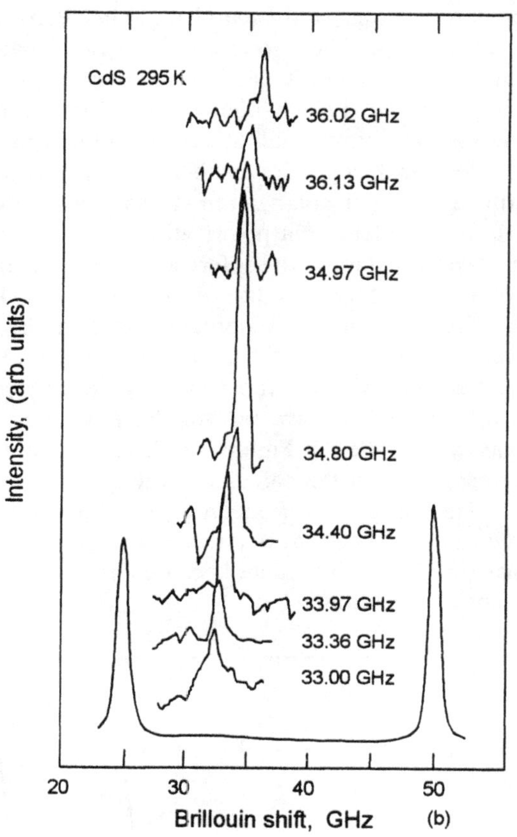

Figure 4.5 (Continued)

An explanation of the experimental result follows. A coherent acoustic wave is present in the crystal, whose acoustic frequency ω_{ac} is tuned around the exact phase matching frequency ω_{aco} with a deviation $\Delta\omega$, as is given by:

$$\omega_L - \omega_s = \omega_{ac} = \omega_{aco} + \Delta\omega$$

$$\Delta q = k_L + k_s - q \tag{4.21}$$

where $\Delta q = -\Delta\omega / v_{ac}$ represents the momentum mismatching. Taking account of Δq, we have the following equation for $E_s(z)$, instead of (4.13):

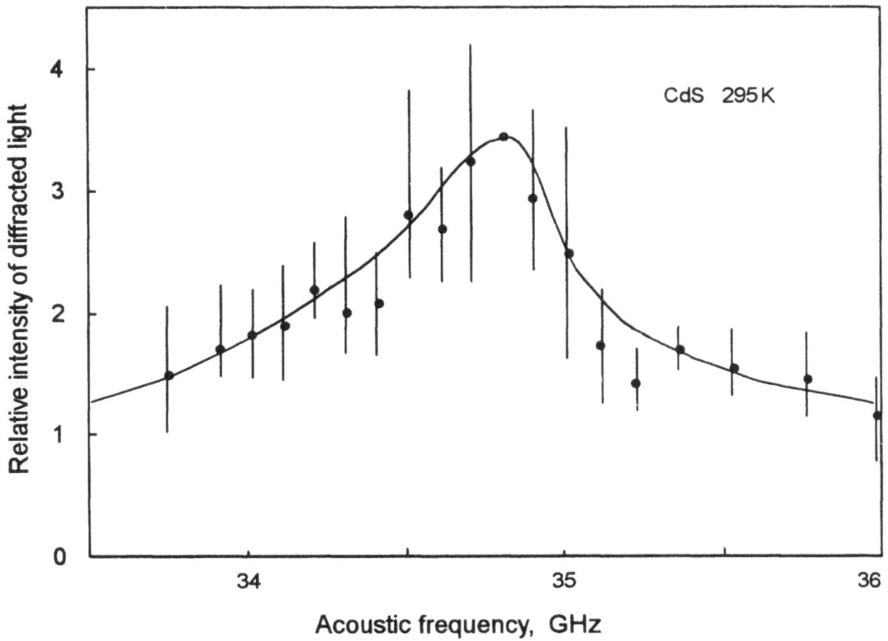

Figure 4.6 Diffracted light intensity corrected for the frequency dependence of the electric field in the cavity. The broken line is a smooth curve drawn through the data points. The error bars represent experimental scatter on subsequent repetition of the experiments [5].

$$\frac{\partial E_s}{\partial z} - j\frac{\Delta q}{2}E_s = -j\frac{k_s}{2}\frac{\varepsilon}{\varepsilon_0}P_{23}E_L S_3^* \tag{4.22}$$

Because the incident light power is weak, the spatial dependence of $S_3(z)$ and $E_s(z)$ can be assumed as:

$$\begin{aligned} E_s(z) &= E_s(0)e^{-\Gamma/2v_{ac}z} \\ S_3(z) &= S_3(0)e^{-\Gamma/2v_{ac}z} \end{aligned} \tag{4.23}$$

Then, from (4.22), we have the relation:

$$\frac{|E_s(0)|^2}{|E_L(0)|^2} = \frac{k_s^2\left(\frac{\varepsilon}{\varepsilon_0}P_{23}\right)^2}{\Gamma^2 + \Delta\omega^2}|S_3(0)|^2 \tag{4.24}$$

Therefore, the backward-diffracted light power is in proportion to the acoustic power, $P_{ac} = C_{33}\, v_{ac}|S_3|^2$, and has a Lorentzian shape around the phase matching frequency ω_{aco} with a width of 2Γ. The full width at half-maximum of the $|E_s(0)|^2/|E_L(0)|^2$ spectrum is found to be 1.4 GHz (i.e., $\Gamma/2\pi$ = 0.7 GHz, from Figure 4.6).

This value corresponds to an attenuation length of $L_{ac} \simeq 1\ \mu$m. The polished crystal surfaces usually contain many defects, which may have increased the damping constant Γ. As will be described in the next section, the damping constant in a silica glass fiber is only 15 MHz, much smaller than those in semiconductors. However, it should be remembered that the value of M_{mn} in semiconductors is larger than that of a silica glass fiber. To achieve a practical semiconductor Brillouin laser, the waveguide structure discussed in Chapter 5 will be necessary.

4.3 FIBER BRILLOUIN LASER

In the early years, the stimulated Brillouin scattering experiments needed very strong pump powers, usually exceeding several tens of megawatts. Moreover, once the stimulated scattering occurred, crystals were often damaged at the input surfaces. The optical fiber has a great advantage in that the interaction length for the nonlinear optical processes, such as the stimulated Brillouin scattering and the stimulated Raman scattering, can be made approximately 10^3 to 10^4 times those in the bulk media.

Ippen and Stolen first found the stimulated Brillouin scattering in the optical fiber at a pump power level less than 1W [6]. Figure 4.7 illustrates their experimental method. The pump beam from a pulsed xenon laser (λ = 5,355 A) is introduced into a single-mode optical fiber with a length of approximately 5m to 20m and a core diameter of 3.8 μm.

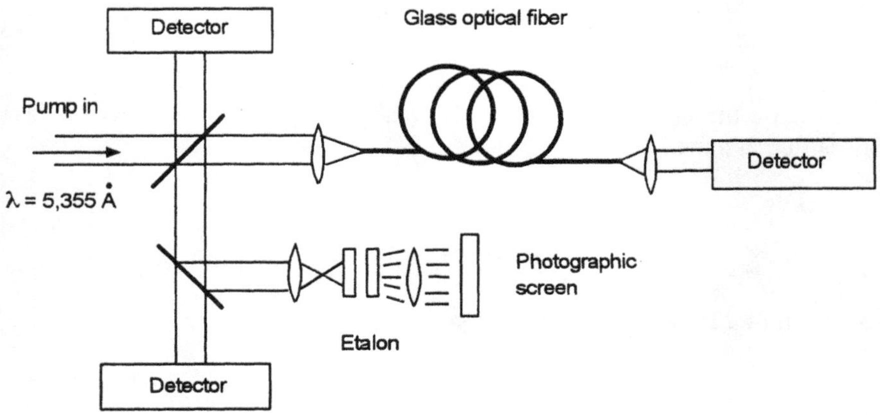

Figure 4.7 Experimental arrangement for observation of backward stimulated Brillouin scattering [6].

The stimulated Brillouin scattering process in the backward direction amplifies the spontaneously emitted Stokes radiation until the pump power depletion becomes significant, as is seen in the oscilloscope traces shown in Figure 4.8. The Brillouin shift was measured as 32.2 GHz, and the gain was $g = 4.3 \times 10^{-9}$ cm/W.

It should be noted that in this experiment by Ippen and Stolen, the xenon laser was equipped with an internal etalon, which kept the laser linewidth narrower than 100 MHz. On the other hand, the Brillouin linewidth in the optical fiber, which corresponds to the damping constant of the acoustic wave, $\Gamma/2\pi$, was estimated to be approximately 100 MHz. In such a situation, the entire pump power can contribute to the amplifying process, so the pump threshold can be quite low. Later, Olsson and Van der Ziel found that the Brillouin gain band in the optical fiber is composed of a main band as narrow as 15 MHz and a bump with a separation of 75 MHz, because of the core-cladding construction [7]. Actually, the pump threshold was reduced to 5 mW or less for a long-length fiber.

For a long-length fiber with a length $L > 1/\alpha$, where α is the optical loss of a fiber, the effective interaction length is α^{-1}. Thus, the gain G is given by:

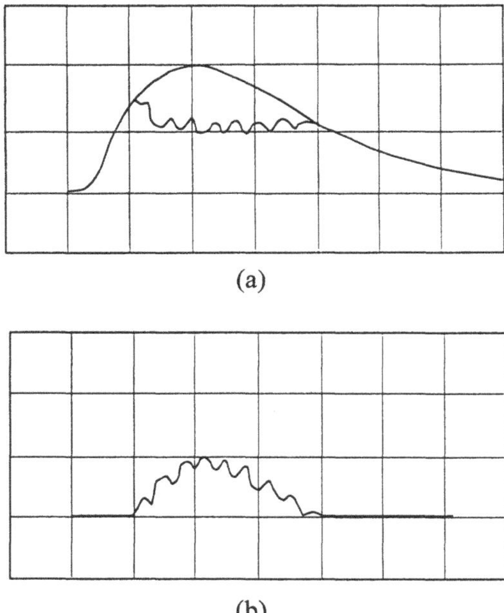

Figure 4.8 Oscilloscope traces of (a) input and transmitted signals and (b) stimulated backward scattering. Fiber length, 5.76 m; time scale, 200 ns/div [6].

$$G = \int_0^L gP_0(\omega_L)\exp(-\alpha z)\,dz \simeq gP_0(\omega_L)\alpha^{-1} \tag{4.25}$$

In this situation, there is no resonance effect by the resonator mirrors as in the conventional laser oscillators. However, spontaneously emitted stokes radiations along the fiber build up to the strong Stokes power at the input end. Smith showed that the Stokes power amplified from the spontaneous emission reaches a magnitude comparable to the pump power above a critical pump power density P_c at which the gain becomes approximately $G = 20$ (i.e., at $P_c \simeq 20\ \alpha/g$) [8]. For example, an optical fiber with $\alpha = 5 \times 10^{-5}$ cm(= 20 dB/km) and the effective core area of $A = 10^{-7}$ cm^2 has a critical pump power estimated to be $AP_c \simeq 20$ mW if the pump laser has a linewidth of approximately 100 MHz.

The low-threshold stimulated Brillouin scattering process cannot occur in usual optical pulse transmissions because the spectrum of a pulse-modulated light wave is broadened much wider than the Brillouin linewidth. Conversely, the stimulated Brillouin scattering is thought to cause a difficulty for a light wave transmission with a small modulation index, such as for cable television signals. In such a case, most of the optical power is concentrated at a carrier wave frequency with a narrow linewidth. Then, the carrier wave is easily converted to the Stokes wave, resulting in the degradation of linearity as well as an increase of the noise figure. The threshold for the stimulated Brillouin scattering for the cable-television-modulated light wave is almost the same as that for the continuous wave (*cw*) light wave [9].

The applicability of the stimulated Brillouin scattering to the light amplification is limited by the narrow amplifier-band nature. Olsson and Van der Ziel showed that the effective amplifier bandwidth can be increased by modulating the frequency of the pump light, although it is important to remember that the gain is inversely reduced [7]. However, they also pointed out that the noise figure of the Brillouin amplifier is quite large. As will be discussed in Chapter 7, the noise figure of the laser amplifier is proportional to the spontaneous emission factor n_{sp}. In the case of the stimulated Brillouin scattering, n_{sp} has the same form as (7.21) for the stimulated Raman scattering, which becomes:

$$n_{sp} = \frac{1}{1 - \exp\left(-\dfrac{\hbar\omega_{ph}}{kT}\right)} \simeq \frac{kT}{\hbar\omega_{ph}} \tag{4.26}$$

because $kT > \hbar\omega_{ph}$ for acoustic phonons. In this case, n_{sp} is then as large as 500 at room temperature. This fact means that the noise figure of the Brillouin amplifier is 500 times that of an ideal amplifier having the same bandwidth.

These natures limit the application of the stimulated Brillouin scattering to the light amplifier. However, as a light wave oscillator, it has the excellent properties

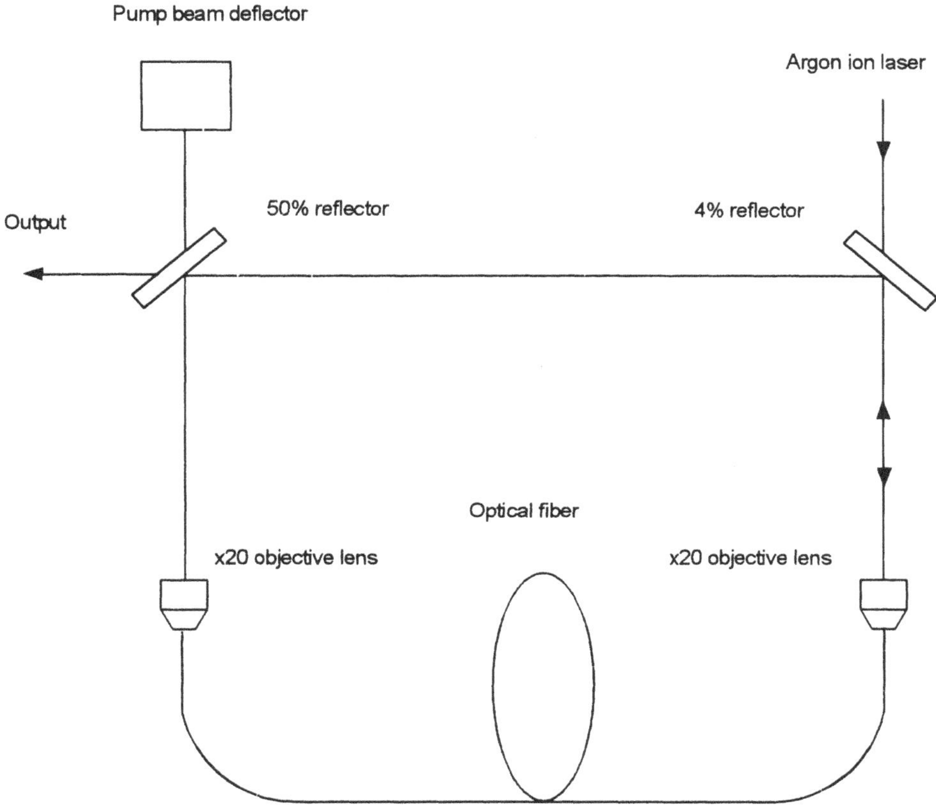

Figure 4.9 Schematic diagram of a fiber Brillouin ring-laser.

of a narrow linewidth and a low-threshold pump power. The Brillouin laser oscillator first reported by Hill and others [10] had a ring resonator arrangement consisting of a fiber 9.5m in length, mirrors, and focussing lenses, all pumped by an argon laser, as illustrated in Figure 4.9. The Brillouin shifted oscillation had a narrow linewidth of approximately 20 MHz, and the threshold pump power was 250 mW. The threshold pump power of the Brillouin laser was reduced to less than 1 mW by means of a high finesse ring resonator, so even low-power laser diodes can be used for pump sources [11]. Besides applications for fiber-Brillouin lasers in optical communication systems, there are also proposals for sensor applications. For example, a frequency change of a fiber Brillouin laser caused by a small mechanical strain applied to the fiber can be easily detected because the laser linewidth is narrow.

REFERENCES

[1] K. Suto, "Semiconductor Raman and Brillouin Lasers" (in Japanese), *Laser Handbook*, ed. by C. Yamanaka, Tokyo: OHMSHA Ltd., Tokyo 1982, Sec.18.5, pp. 285–290.

[2] K. Okamoto, H. Ishikawa and J. Nishizawa, "Ultrasonic Wave Generation in Oscillating *n*-GaAs," *RIEC Technical Report*, Tohoku Univ., TR22, 1967, pp. 1–8.

[3] J.P. Laurenti, "Dispersion of the P_{44} Photoelastic Coefficient in CdS at Liquid Nitrogen Temperature," *Solid State Commun.*, Vol. 41, 1982, pp. 177–179.

[4] C.F. Quate, C.D.W. Wilkinson and D.K. Winslow, "Interaction of Light and Microwave Sound," *IEEE Proc.*, Vol. 53, 1965, pp. 1604–1623.

[5] S. Suzuki, J. Nishizawa and K. Suto, "The Coherent Interaction of Externally Generated 35-GHz Sound with the Light in CdS," *Appl. Phys. Lett.*, Vol. 30, 1977, pp. 310–312.

[6] E.P. Ippen and R.H. Stolen, "Stimulated Brillouin Scattering in Optical Fiber," *Appl. Phys. Lett.*, Vol. 21, 1972, pp. 539–541.

[7] N.A. Olsson and J.P. Van der Ziel, "Characteristics of a Semiconductor Laser Pumped Brillouin Amplifier with Electronically Controlled Bandwidth," *J. Lightwave Technol.*, Vol. LT5, 1987, pp. 147–153.

[8] R.G. Smith, "Optical Power Handling Capacity of Low Loss Optical Fibers as Determined by Stimulated Raman and Brillouin Scattering," *Appl. Optics*, Vol. 11, 1972, pp. 2489–2494.

[9] X.P. Mao, G.E. Bodeep, R.W. Tkach, A.R. Chraplyvy, T.E. Darcie and R.M. Derosier, "Brillouin Scattering in Externally Modulated Lightwave AM-VSB CATV Transmission Systems," *IEEE Photon. Technol. Lett.*, Vol. 4, 1992, pp. 287–289.

[10] K.O. Hill, B.S. Kawasaki and D.C. Johnson, "CW Brillouin Laser," *Appl. Phys. Lett.*, Vol. 28, 1976, pp. 608–609.

[11] P. Bayvel and I.P. Giles, "Linewidth Narrowing in Semiconductor Laser Pumped All-Fiber Brillouin Ring Laser," *Electron. Lett.*, Vol. 25, 1989, pp. 260–262.

Chapter 5
Heterostructure Raman Laser

5.1 FUNDAMENTALS OF THE HETEROSTRUCTURE RAMAN LASER

5.1.1 GaP—$Al_xGa_{1-x}P$ Heterostructure

Light waves can be confined in a narrow layer if the layer is sandwiched by other layers with a lower refractive index. Although there are many heterostructure combinations that give such optical confinement, the other important consideration is lattice matching at the heterostructure interface. If the lattice mismatching is too large, a great number of imperfections are produced at the interface, so the light is strongly absorbed or scattered. Table 5.1 gives the lattice parameter mismatching, $\Delta a/a$, for various combinations of III–V compound semiconductors.

It should be noted that the difference in the lattice constants of $Al_xGa_{1-x}P$ and GaP is as small as that between $Al_xGa_{1-x}As$ and GaAs. As shown in Figure 5.1, we made an X-ray measurement of the lattice constant difference on $Al_xGa_{1-x}P$ layers epitaxially grown on GaP substrates and found that $(a_{AlP} - a_{GaP})/a_{GaP} = 3.5 \times 10^{-3}$ [1]. This value is comparable to the value in the AlAs/GaAs system (i.e., $(a_{AlAs} - a_{GaAs})/a_{GaAs} = 1.6 \times 10^{-3}$). Nevertheless, this small difference in the lattice constants seriously affects the waveguiding characteristics in the heterostructure due to strain-induced optical anisotropy. This subject will be discussed later.

The refractive indices of various III–V and II–VI compounds are summarized in Figure 5.2 as a function of wavelength [2]. To a first approximation, we can assume that the refractive index of an alloy of III–V compounds linearly depends on the alloy composition x. Then, the room temperature refractive index of $Al_xGa_{1-x}P$ at wavelength approximately = 1.0 μm is given by:

$$n_{Ax} = 3.218 - 0.343x \qquad (5.1)$$

Table 5.1
Lattice Mismatching in III-V Compound Semiconductors

Epitaxial Layer	*Substrate*	*Maximum* $\Delta a/a$
GaAlAs	GaAs	1.6×10^{-3}
GaAlP	GaP	2.2×10^{-3}
GaPAs	GaAs	3.6×10^{-2}
InPAs	InP	3.1×10^{-2}
AlPAs	AlAs	3.5×10^{-2}
InGaP	GaP	7.0×10^{-2}
InGaAs	GaAs	6.7×10^{-2}
InAlAs	InAs	6.7×10^{-2}
InAlP	InP	6.9×10^{-2}

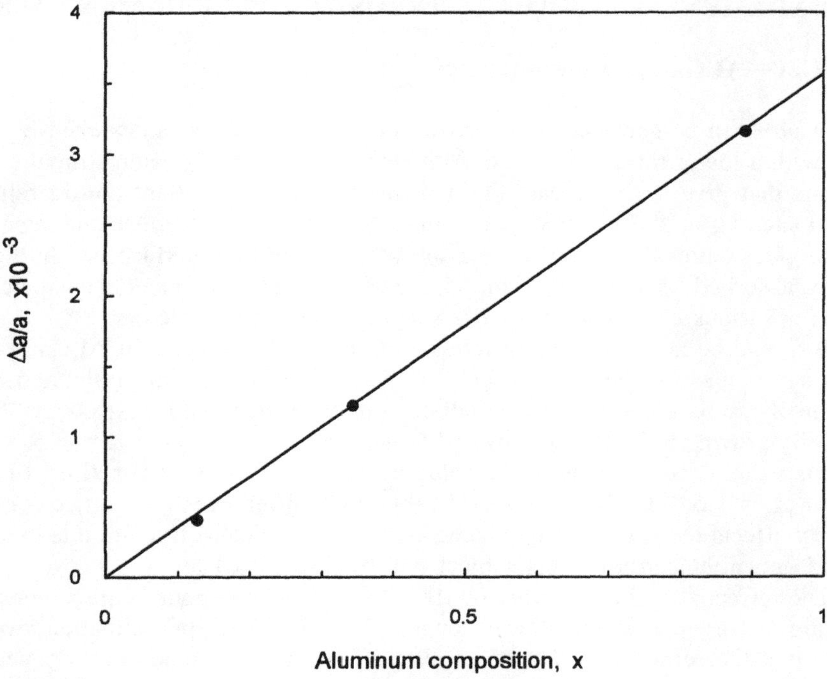

Figure 5.1 Mismatching of lattice parameters, $\{a(Al_xGa_{1-x}P)\text{-}a(GaP)\}/a(GaP)$, as a function of aluminum composition x [1].

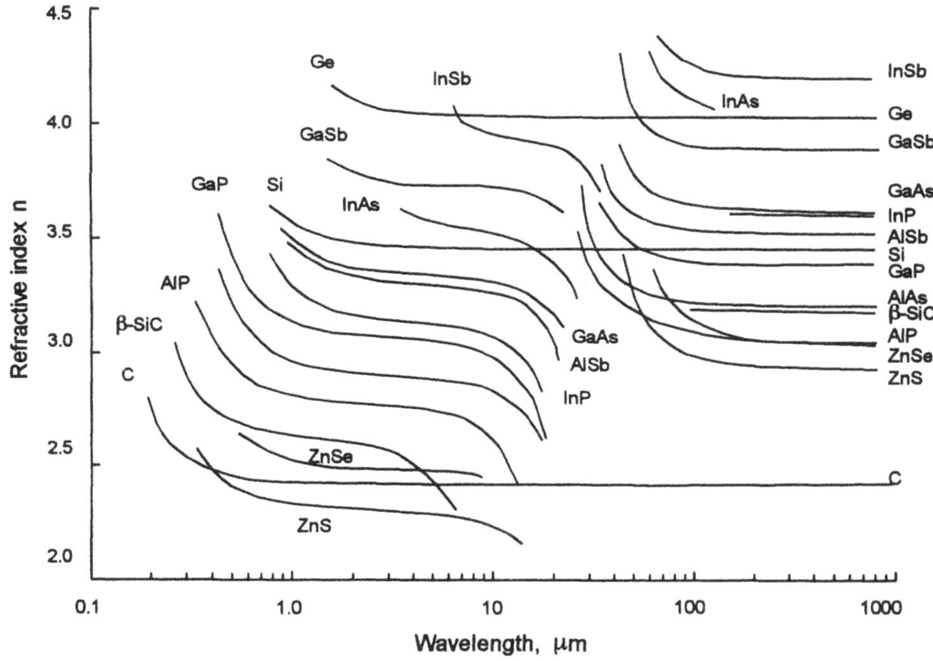

Figure 5.2 Spectral dependences of the refractive indices of IV, III–V, and II–VI semiconductor at $T = 20°C$ [2].

Although we will later discuss the guided wave, we now roughly estimate how thin a layer can be within which the light wave can be confined, due to the refractive-index difference. This is illustrated in the inset of Figure 5.3.

Total reflection of a plane wave incident at the interface of the heterostructure occurs when the incident angle is less than the critical angle θ_c, given by:

$$\sin \theta_c = \sqrt{\frac{2\Delta n}{n_1}\left(1 - \frac{\Delta n}{2n_1}\right)} \tag{5.2}$$

where n_1 and $n_2 = n_1 - \Delta n$ are the refractive indices of the adjacent layers.

The thickness of the layer that gives substantial optical confinement is given by:

$$d_c = \frac{\lambda_0}{2n_1 \sin\theta_c} \tag{5.3}$$

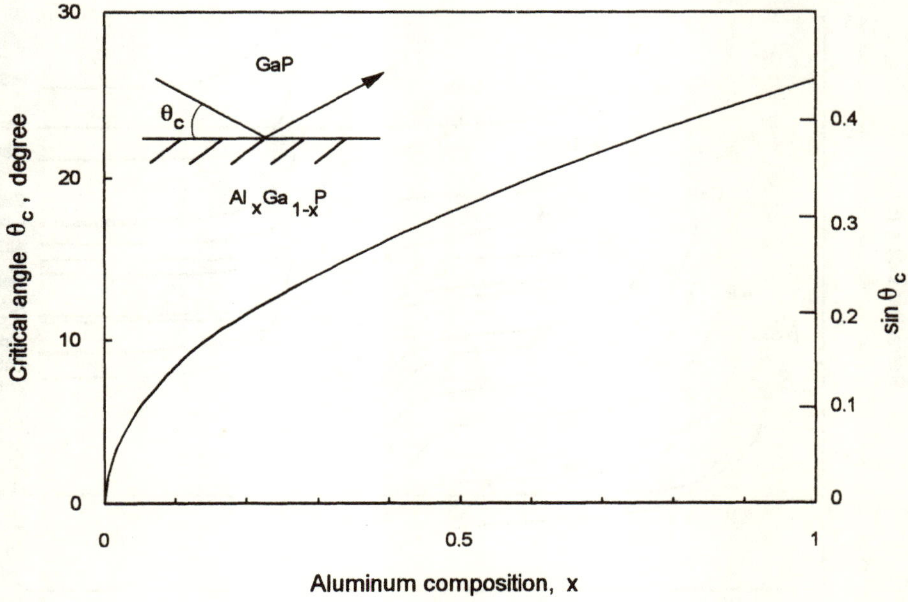

Figure 5.3 Critical angle θ_c for total reflection at the GaP/$Al_xGa_{1-x}P$ interface and $\sin\theta_c$ as a function of aluminum composition x [1].

If the thickness d is smaller than this value, the optical field spreads considerably into the cladding layers. Figure 5.3 shows the calculated $\sin\theta_c$ as a function of the aluminum composition x in $Al_xGa_{1-x}P$. For example, at $\lambda_0 = 1.0$ μm and $x = 0.1$, we obtain $d_c = 1.07$ μm.

5.1.2 STRAIN-INDUCED OPTICAL ANISOTROPY

The waveguiding characteristics of the heterostructure Raman laser are affected by the optical anisotropy induced by the lattice strain due to the small difference in the lattice constants of $Al_xGa_{1-x}P$ and GaP [3,4]. It is observed that a He-Ne laser beam with $\lambda = 0.63$ μm introduced into the substrate region emerges from the output endface at a small angle θ_h relative to the axis, as illustrated in Figure 5.4, if the polarization is parallel (i.e., horizontal) to the epitaxial-layer surface on a (100)-oriented substrate crystal. On the other hand, the deviation of the beam direction is too small to be observed if the polarization is perpendicular to the epitaxial-layer surface. The angle θ_h for the horizontal polarization beam ranged from 2 to 5 mrad, increasing with the thickness of the $Al_xGa_{1-x}P$ layers and with the value of x. This

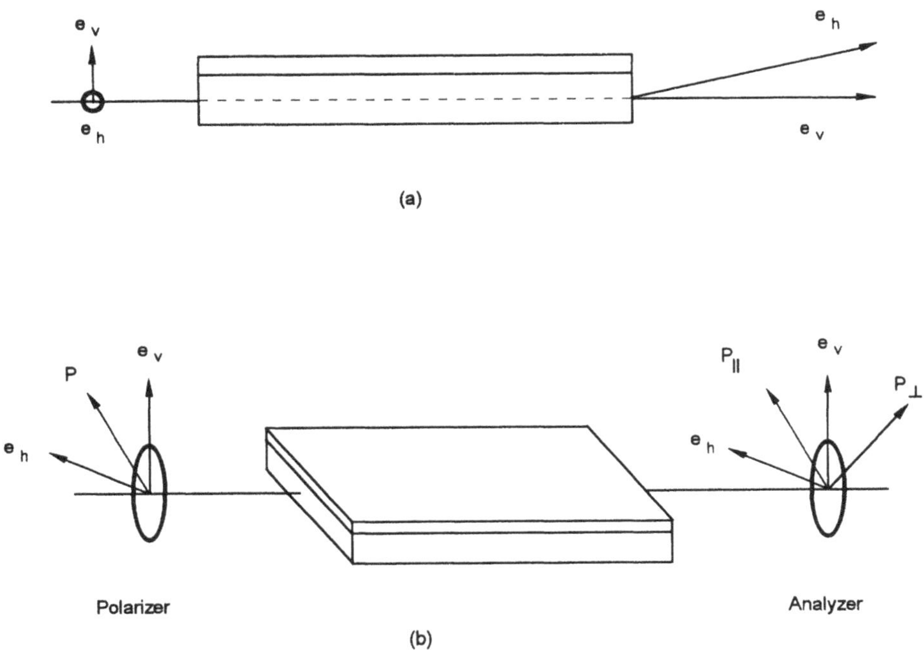

Figure 5.4 (a) Beam deflection due to the inhomogeneous optical anisotropy. (b) Experimental arrangement for the measurement of optical anisotropy. Polarization vector **P** of the polarizer is 45° from both $\mathbf{e}_h$ and $\mathbf{e}_v$ [3].

effect is due to the gradual increase of the refractive index n_h for the horizontal polarization from the bottom surface of the substrate to the epitaxial-layer surface caused by the curvature of the wafer.

In practice, the two resonator faces were adjusted to be optically parallel for the horizontal polarization, rather than geometrically parallel, by observing a He-Ne laser beam with the horizontal polarization reflected back at the output face. Our YAG laser for pumping is usually vertically polarized so that Stokes radiation with horizontal polarization is pumped according to the selection rule. The distribution of the optical anisotropy in the substrate was measured by the experimental configuration shown in Figure 5.4. The polarization direction of the polarizer is 45 deg from the optic axes (vertical and horizontal). An incident He-Ne laser beam with a diameter of 100 μm at a polarization direction of 45 deg is shone through the substrate crystal at a height d from the bottom surface of the substrate. The transmitted light intensity, $I_{\parallel}$ or $I_{\perp}$, is measured by making the polarization direction of the analyzer parallel or perpendicular, respectively, to the polarization direction of the polarizer.

Optical anisotropy is given by the following equation:

$$\frac{I_\perp}{I_\parallel + I_\perp} = \sin^2 \frac{\delta}{2} \tag{5.4}$$

where δ is the phase shift given by:

$$\delta = \Delta n = \frac{2\pi}{\lambda} l \quad \Delta n = n_h - n_v \tag{5.5}$$

where n_h and n_v are the refractive indices for the horizontal and vertical polarization, respectively. The length of the light path, l, is 3.5 mm for typical heterostructure Raman lasers. Figure 5.5 shows the results for several heterostructure wafers for Raman lasers without lateral optical confinement. (See Section 5.1.3.)

For some samples, $I_\perp/(I_\parallel + I_\perp)$ almost reaches 1.0, corresponding to a phase shift of π. The refractive-index difference giving $I_\perp/(I_\parallel + I_\perp) = 1.0$ is $\Delta n = 0.9 \times 10^{-4}$, when $l = 3.5$ mm.

Figure 5.5 shows that the refractive-index difference increases with increasing d, the height from the bottom surface of the substrate. This result indicates that the strain in the GaP substrate is highest at the upper surface and gradually decreases along the thickness direction of the substrate over a 200 μm distance. Therefore, we can expect that the index change in the Raman active layer is of the same order of magnitude as that in the uppermost region of the substrate.

Although the probing beam diameter is not narrow enough, Figure 5.5 gives no apparent indication of the compressive strain region near the bottom side. The reason is not yet clear, but it may be related to the fact that the bottom side has a rough surface that contains many dislocations and can absorb the strain by annealing during the growth process. On the other hand, samples with greater mismatch do show the existence of a compressive strain region near the bottom surface, as depicted in Figure 5.6.

The changes in the refractive indices for horizontal and vertical polarization are given by:

$$\Delta n_h = \frac{n^3}{2} \{p_{11} S_{xx} + p_{12}(S_{yy} + S_{zz})\} \tag{5.6}$$

$$\Delta n_v = \frac{n^3}{2} \{p_{12}(S_{xx} + S_{yy}) + p_{11} S_{zz}\}$$

where p_{ij} and S_{ij} are photoelastic constants and strain tensors for GaP. Instead of obtaining accurate strain relations, we can make a rough estimate by simply assuming

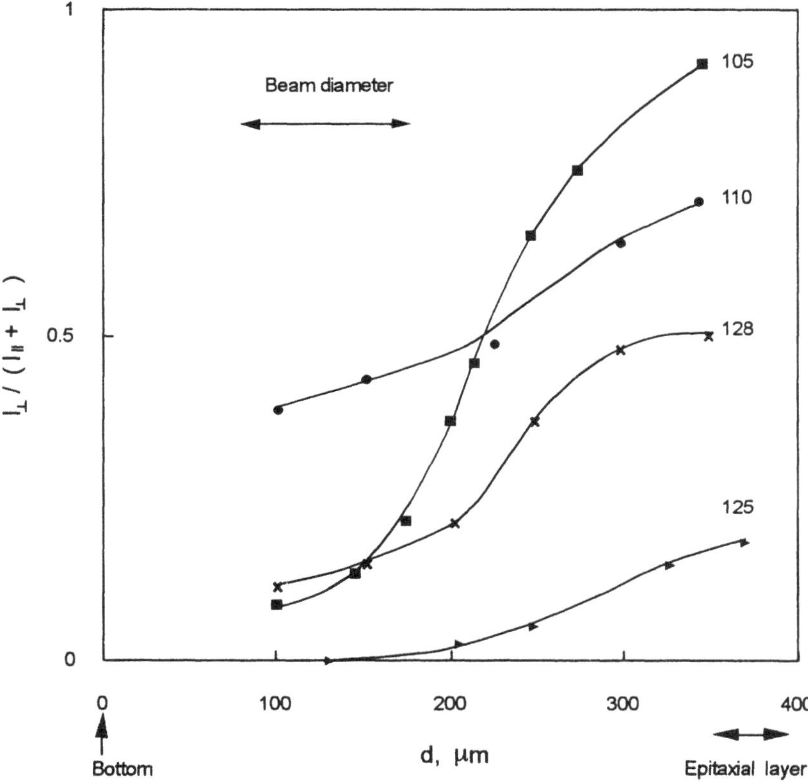

Figure 5.5 Polarization ratios $I_\perp/(I_\parallel + I_\perp)$ as a function of d, the height from the bottom of the substrate. Substrate thickness is 380 μm. Samples 105 and 110 are three-layer structure Raman lasers. Samples 128 and 125 are four-layer structure Raman lasers [3].

that the stress in the z-direction, T_{zz} (vertical to the epitaxial surface), can be neglected compared to the stresses T_{xx} and T_{yy}, near the surface.

Then:

$$T_{zz} = C_{11}S_{zz} + C_{12}S_{xx} + C_{12}S_{yy} = 0 \quad \text{and} \quad S_{xx} \simeq S_{yy}.$$

We get:

$$S_{zz} \simeq -\frac{2C_{12}}{C_{11}}S_{xx}$$

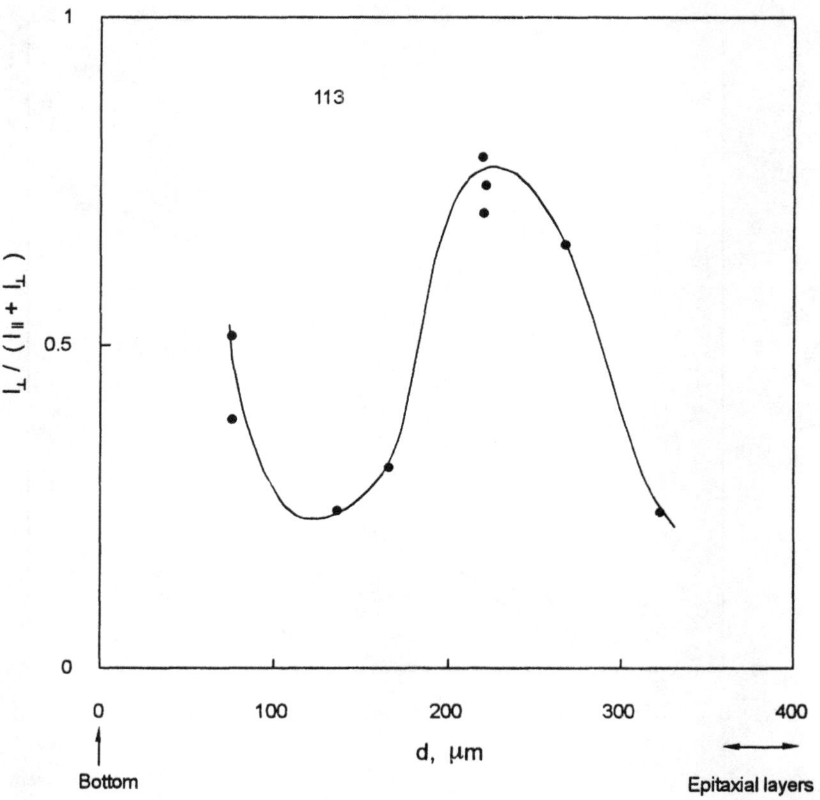

Figure 5.6 Polarization ratio $I_\perp/(I_\parallel + I_\perp)$ as a function of d, the height from the bottom of the substrate for sample 113 [3].

Therefore, n_h and n_v are given by:

$$\Delta n_h \simeq \frac{n^3}{2}\left\{(p_{11} + p_{12}) - \frac{2C_{12}}{C_{11}} p_{12}\right\} S_{xx}$$

$$\Delta n_v \simeq \frac{n^3}{2}\left\{2p_{12} - \frac{2C_{12}}{C_{11}} p_{11}\right\} S_{xx}$$

The photoelastic constants of GaP are known to be $p_{11} = -0.151$, $p_{12} = -0.082$, and $p_{44} = -0.074$ at 0.63 μm, while it is known that $(2C_{12}/C_{11}) = 0.89$.

Then, we get:

$$\Delta n_h = -0.16 \times \frac{n^3}{2} S_{xx}$$

$$\Delta n_v = -0.03 \times \frac{n^3}{2} S_{xx} \tag{5.7}$$

where S_{xx} is the strain in the $\langle 100 \rangle$ direction parallel to the epitaxial surface (i.e., the direction of the resonator axis). Although the length and the width of a chip is slightly different (3.5 and 5.5 mm, respectively), we assume $S_{xx} \simeq S_{yy}$ for simplicity.

The above relation, which shows Δn_h is approximately 5 times Δn_v, well explains the experimental fact that only the beam with horizontal polarization is observed to change its direction. This change is due to the change in S_{xx} along the thickness direction z. It gradually increases with d, as is clearly shown in Figure 5.5.

5.1.3 Intrinsic Optical Anisotropy in GaP

The resonator axis of the semiconductor Raman laser is usually along the $\langle 100 \rangle$ direction. However, we must take into account another kind of optical anisotropy that appears without strain if the resonator axis is chosen to be different from one of the $\langle 100 \rangle$ directions.

For the beam propagating in one of the $\langle 110 \rangle$ directions, the refractive indices $n_{\langle 110 \rangle}$ and $n_{\langle 001 \rangle}$ for the optical polarization directions $\langle 110 \rangle$ and $\langle 001 \rangle$, respectively, are different [5].

In a strain-free homogeneous GaP crystal, optical anisotropy is observed by a measurement system similar to that shown in Figure 5.4, if the beam direction is at an angle from the $\langle 100 \rangle$ direction in the (001) plane.

The results are shown in Figure 5.7, and can be fitted by [3]:

$$\Delta n(\phi) = \Delta n \sin^2 2\phi \tag{5.8}$$

where is the angle between the $\langle 100 \rangle$ axis and the light beam direction, and $|\Delta n| = |\Delta n_{\langle 110 \rangle} - n_{\langle 001 \rangle}| \simeq 1.6 \times 10^{-5}$. This value cannot generally be neglected compared to the strain-induced optical anisotropy when $\phi \neq 0$.

More generally, this anisotropy can be described by an index quasi-ellipsoid with cubic symmetry:

$$\frac{x^2 + y^2 + z^2}{n^2} + \alpha(x^2y^2 + y^2z^2 + z^2x^2) = 1 \qquad \alpha = -\frac{8\Delta n}{n^5} \tag{5.9}$$

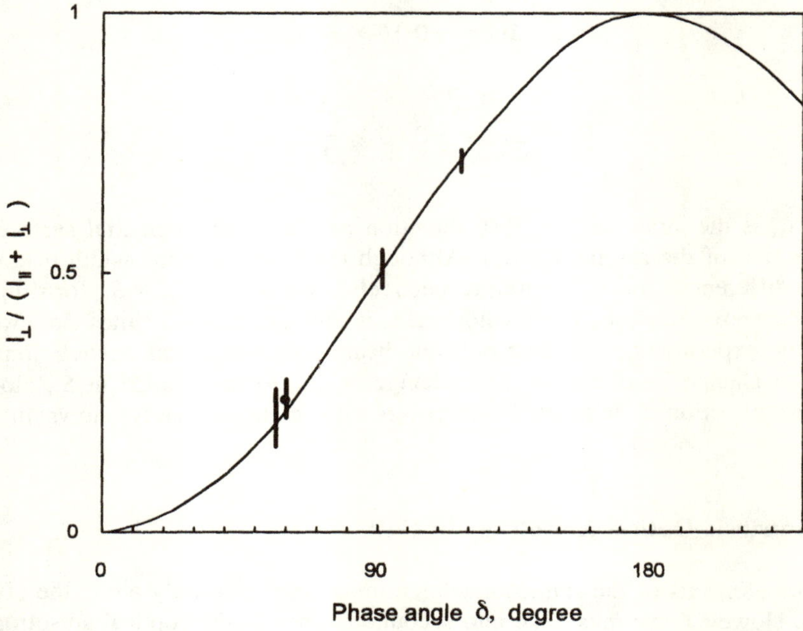

Figure 5.7 Experimental plots of the polarization ratio $I_\perp/(I_\parallel + I_\perp)$ as a function of phase angle δ [3]. $\delta = \Delta n(\psi)\dfrac{2\pi}{\lambda}L$ where L is the length of a crystal. Solid curve represents $I_\perp/(I_| + I_\perp) = \sin^2\dfrac{\delta}{2}$.

where x, y, and z are the cubic axes. It can be shown that the long and short axes of the quasi-ellipse made by cutting the ellipsoid with the plane perpendicular to the beam direction gives the index-difference (5.8).

5.1.4 Raman Oscillation Characteristics of Heteroepitaxial Layers

Figure 5.8 shows the heterostructure Raman laser without any lateral optical confinement, in which an unfocussed YAG laser beam is introduced at the side of the high-reflection film, the same as in the case of the homoepitaxial Raman laser described in Section 3.3. In the later two sections, we describe buried heterostructure Raman lasers in which the GaP core is surrounded by $Al_xGa_{1-x}P$ cladding layers.

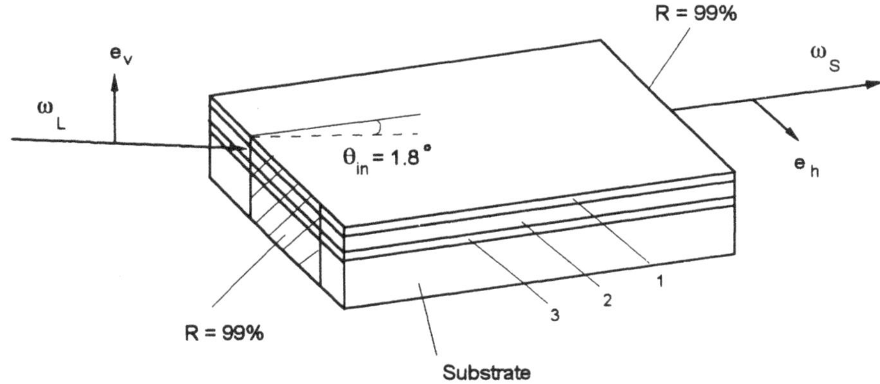

Figure 5.8 Heterostructure semiconductor Raman laser without lateral optical confinement [3]. (1) First $Al_xGa_{1-x}P$ cladding layer; (2) GaP Raman active layer; (3) Second $Al_xGa_{1-x}P$ cladding layer; $\mathbf{e}_h$, $\mathbf{e}_v$ = Horizontal and vertical polarization vectors relative to epitaxial planes.

In Section 5.3, a small window is opened at a part of the cross section of the GaP core region using photolithography. However, in a more promising method discussed in Section 5.4; the input resonator film is designed to be transparent at the pump wavelength but highly reflective at the Stokes light wavelength. In the latter method, the cross section of the waveguide can be made as small as the focussed spot size of the pump beam. The waveguide can be made more narrow by introducing a tapered waveguide structure, as will be described in Section 5.5. In this section, however, we describe the lasing characteristics of the broad heterostructure illustrated in Figure 5.8 [3], which should be useful in understanding the general features of Raman oscillation characteristics in heterostructures. The beam diameter of the pump YAG laser is as large as 6 mm so that only a part of the pump beam is introduced into the resonator. Therefore, the input power level is described as the power density.

Figures 5.9 and 5.10 show the lasing threshold characteristics of three-layer structure Raman lasers with aluminum compositions x of the cladding layers around 0.1. The thickness of each layer is shown in the insets of the figures. The lasing threshold pump power density (2.5×10^6 W/cm^2) for sample number 110 with a Raman active-layer thickness of 15 μm is as low as those of the homoepitaxial GaP Raman lasers with much thicker active layers (60 ~ 90 μm) described in Section 3.3.

However, it should be noted that the optical anisotropy is not small even when $x \simeq 0.1$ if the second cladding layer is thick (see Figure 5.5).

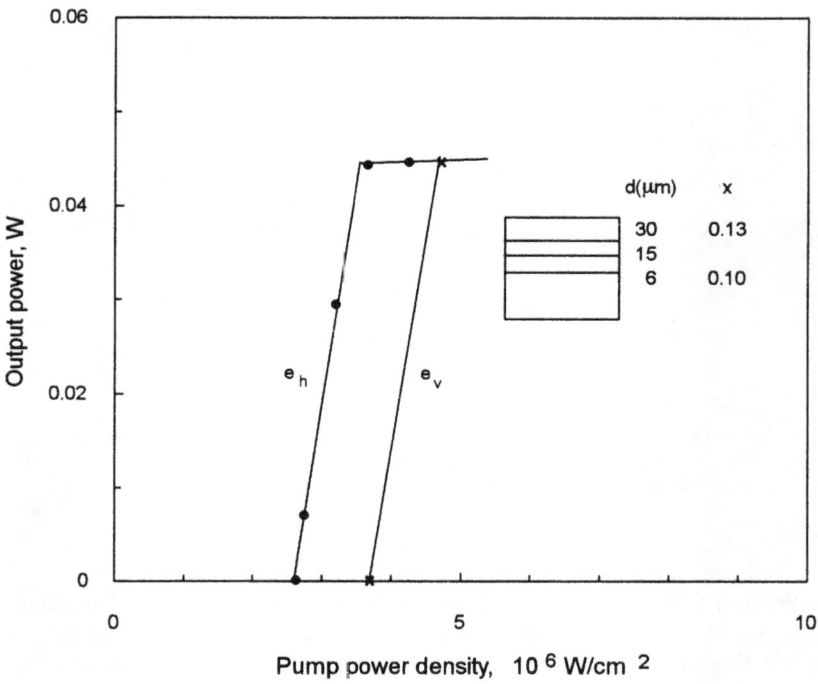

Figure 5.9 Output power of the Stokes radiation as a function of pump power density for a three-layer Raman laser (number 110) with a Raman active layer thickness 15 μm. In the figure, $\mathbf{e}_h$ means that the polarization of the Stokes radiation is $\mathbf{e}_h$ and that of the pump radiation is $\mathbf{e}_v$, while $\mathbf{e}_v$ means the reverse [3].

Optical anisotropy causes a change in the direction of the light beam having horizontal polarization.

The lasing threshold is slightly lower when the pump field is vertically polarized and the resultant Stokes field has a horizontal polarization than when the light fields have the opposite polarizations.

This is because the resonator was adjusted to be parallel for the horizontally polarized beam. We have examined cladding layers with higher aluminum compositions ($x \simeq 0.2 \sim 0.3$). As shown in Figure 5.10, the thicknesses of the first and second cladding layers of sample number 113 are 2 μm and 8 μm, respectively. Although the cladding layers are thinner in this sample than in number 110, the optical anisotropy is very large, as was shown in Figure 5.6. As discussed earlier, the phase difference between the beams with polarization directions parallel and perpendicular to the epitaxial plane is as much as π for this sample (see Figure 5.6).

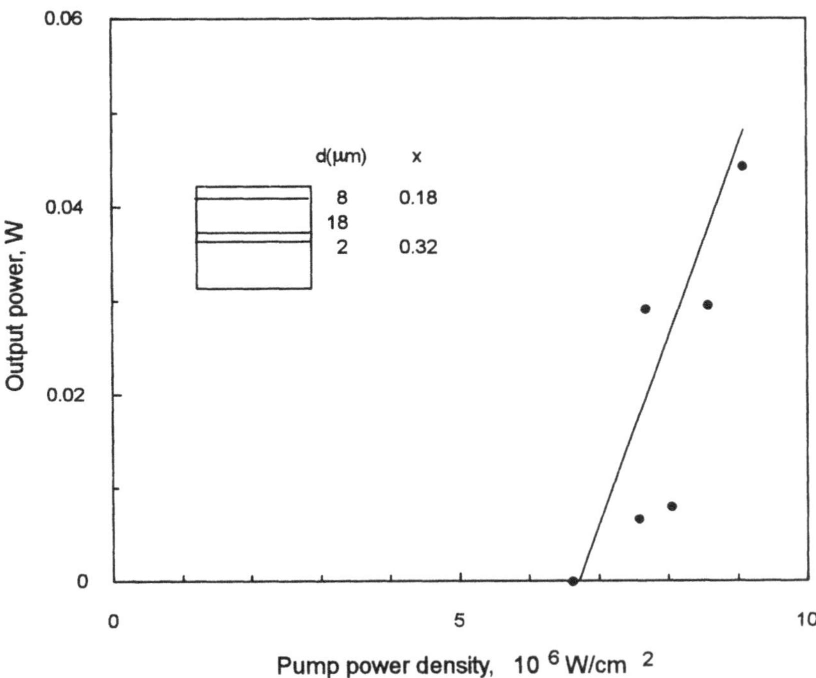

Figure 5.10 Output power of the Stokes radiation as a function of pump power density for a three-layer (number 113) with higher aluminum compositions of the cladding layers [3].

This large optical anisotropy may be one of the causes of the high-threshold characteristics. Also, the resonator parallelism may be poorer in this sample because the thickness of the second cladding layer is only 8 μm; that is, a mechanically polished surface is not perfectly flat but has considerable curvature at the near edge regions with width wider than 10 μm.

To be able to use thinner cladding layers without losing the resonator parallelism, we have fabricated the four-layer structure shown in Figure 5.11. In this structure, another GaP layer (layer 4) with thickness 10 to 20 μm is grown over the thin second cladding layer (layer 3).

Figure 5.11 shows the lasing threshold characteristics for one of the four-layer Raman lasers. The threshold optical power density is reduced to 4×10^6 W/cm^2, although the x values in the cladding layers are as large as approximately 0.26 to 0.27. The phase difference due to optical anisotropy in this sample is reduced to approximately $1/2\ \pi$ (see Figure 5.5).

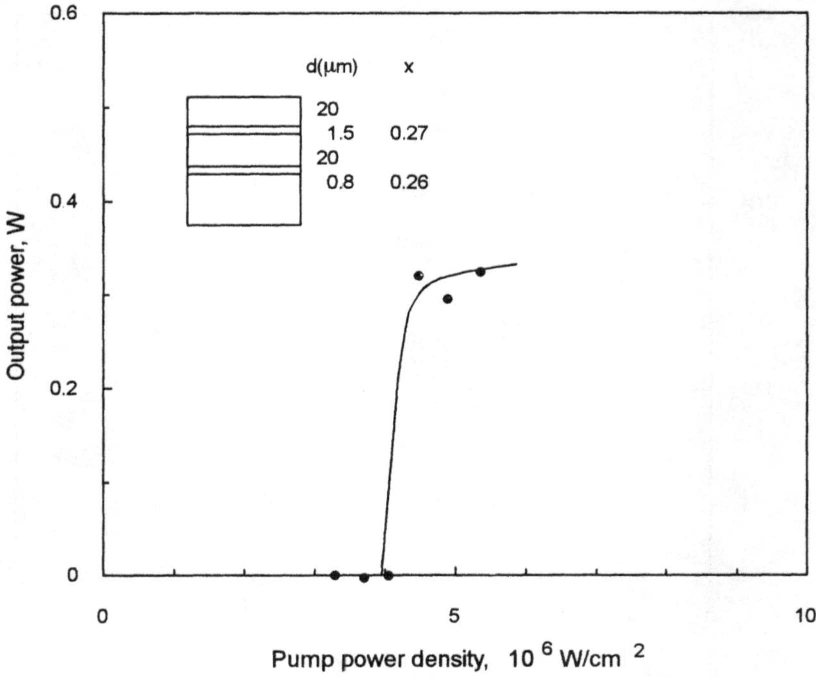

Figure 5.11 Output power of the Stokes radiation as a function of pump power density for a four-layer Raman laser (number 128) [3].

5.2 BURIED WAVEGUIDE RAMAN LASER WITH MULTIMODE PUMPING

5.2.1 Waveguide Fabrication

In comparison to the single-mode pumped buried waveguide semiconductor Raman laser, which will be described in Section 5.3, the multimode pumped Raman laser described here needs a relatively high pump power. However, it has an advantage in that it can oscillate over a wide range of pump wavelengths without changing the resonator films. Figure 5.12 shows the structure of the multimode pumped waveguide [6,7]. It has a window opened in part of the input resonator film. The pump light introduced from the narrow window spreads over the entire waveguide width and is multiply reflected at both ends by the reflection film. As a result, the pump field distribution is expected to be multimoded in the waveguide.

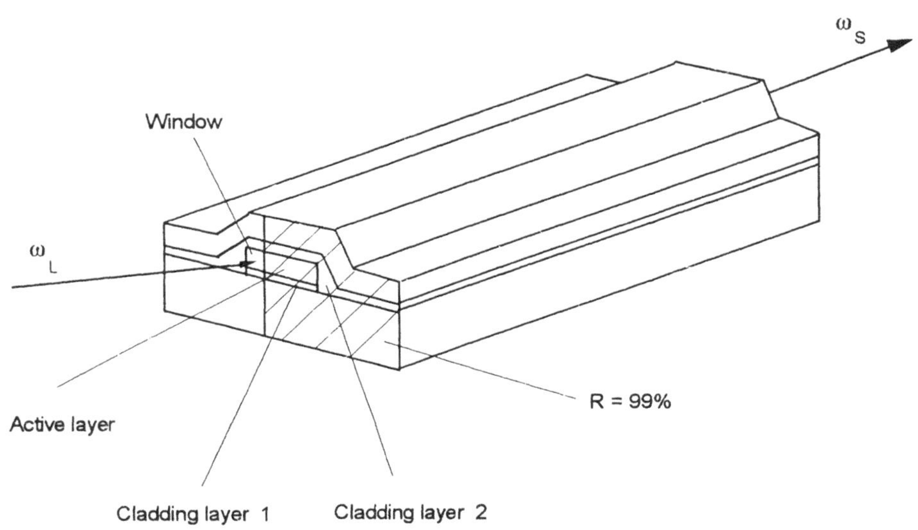

Figure 5.12 Buried heterostructure semiconductor Raman laser with an input window.

The fabrication procedure of the semiconductor waveguide and resonator film structure is as follows:

1. The first cladding layer of $Al_{0.1}Ga_{0.9}P$ (thickness $d \sim 0.5$ to 2 μm) and the GaP Raman active layer ($d \sim 5$ to 10 μm) are successively grown on a GaP substrate having a (100) orientation.
2. Stripes in the epitaxial regions with widths of 30 and 40 μm are formed by the plasma etching technique. We find a high etch rate and smooth surface of GaP is obtained using PCl_3 gas without seriously affecting the photoresist mask.

   PCl_3 gas is introduced into a plasma chamber with parallel plate electrodes at a pressure of 0.05 Torr and is excited by an RF generator operating at 3.2 MHz. A sample crystal is placed on a quartz pedestal placed on a ground electrode. Etching rates as high as 1.0 μm/min are obtained at an RF power of approximately 200 W, and the etched side walls are nearly perpendicular to the epitaxial plane. Figure 5.13 shows a SEM photograph of the etched stripe.
3. Then, the upper cladding layer of $Al_{0.1}Ga_{0.9}P$ (the third layer, $d \sim 1$ to 5 μm) and the GaP buffer layer (the fourth layer, $d \sim 10$ to 20 μm) are successively grown.
4. The two end-faces with (100) orientation are polished to an optical flatness and chemically etched to remove the resulting heavily damaged surface layer.

Then, 99% reflection films made of alternating 13 layers of SiO_2 and TiO_2, each $\lambda/4$ thick, are evaporated on them.

5. The window for the pump beam is opened through one of the reflection films by the photolithographic technique. The thin platelet is sandwiched between a pair of glass plates, and positive photoresist is coated on one of its end-faces. The reflection film at the window region is then removed by dipping in an etching solution of $HF:H_2O$ = 1:15 for 30 sec.

Windows with a width of approximately 3 to 7 μm are opened for the stripes approximately 30 to 40 μm in width.

There are some problems in the fabrication of low-threshold buried-heterostructure Raman lasers. Smaller stripes tend to cause extra transmission loss due to light scattering at the interfaces in the lateral directions of the stripe because of interfacial irregularities. Morphological imperfections such as gallium occlusions are often a problem because the laser length is as long as approximately 3 to 5 mm. Narrow windows with width of approximately 3 to 7 μm must be opened, but the inevitable lateral etching of the high-reflection films can easily be more than 3 μm because they are composed of 13 layers of SiO_2 and TiO_2 with total thicknesses of approximately 1.5 μm. Figure 5.14 shows the photographs of the input surfaces of heterostructures with the high-reflection films partially opened for the windows.

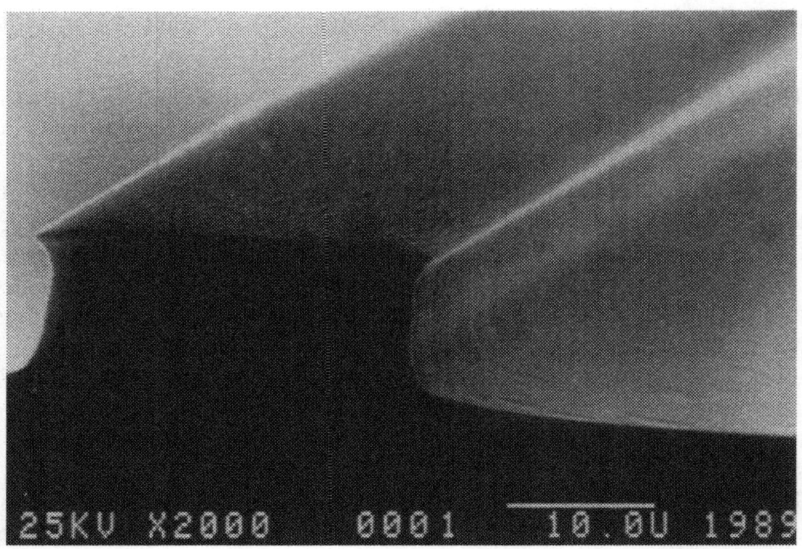

Figure 5.13 Scanning electron micrograph (SEM) of a stripe of the first cladding layer and the active layer formed by the RIE process [7].

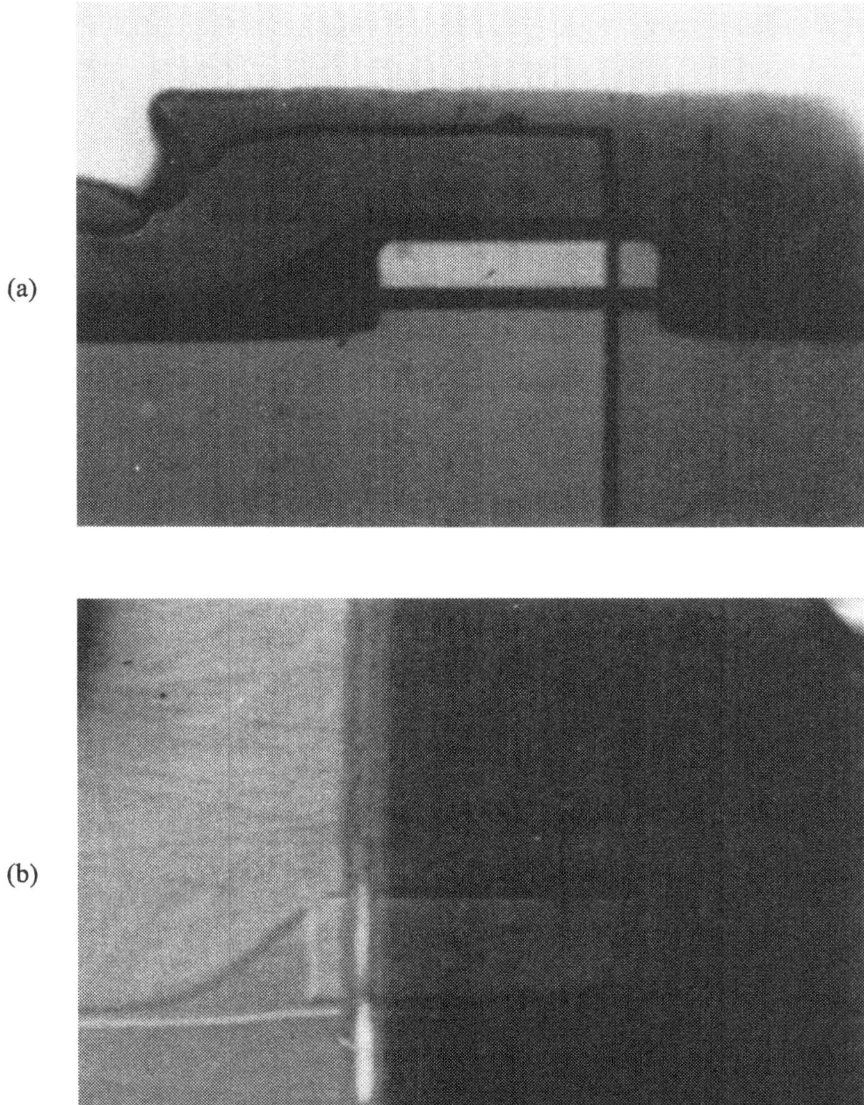

Figure 5.14 Photographs of the input sides of two different heterostructure semiconductor Raman lasers. The etching depth is deeper in (a) than in (b). Active layer thicknesses are 9 μm in both cases. The stripe in (a) is illuminated from the output side to enhance the contrast. In the case of (b), it is seen that acute angles appear at the corners of the side walls [7].

The optical transmission of these stripes had been measured before deposition of the high-reflection films, so the actual optical losses of the waveguide could be known. This was performed by using a *cw* dye laser at a wavelength of 600 to 620 nm, rather than at the lasing wavelength of 1.11 μm, because *cw* laser light is necessary for precise optical transmission measurements. The beam from the dye laser was focussed through a lens with f = 10 mm.

Figure 5.15 shows the results for all the stripes on a single wafer. After correction for the multiple reflections at both ends, the transmission was found to be 90% to 95%. The effect of light scattering and leakage at the hetero-interface can be estimated by subtracting the internal absorption loss from the above values. We have measured the absorption coefficient at λ = 600 ~ 630 nm by growing thick undoped GaP layers and obtained $\alpha \simeq 0.13 \sim 0.15$ cm^{-1}. This value of corresponds to a transmittance of 95% to 96%, so the loss due to interface scattering and leakage is estimated to be 5% or less.

The unintentionally doped GaP layers are usually p-type, with carrier concentration $p \simeq 1.5 \sim 2 \times 10^{16}$ cm^{-3} for growth temperatures of approximately 880 to 870°C. However, the conductivity tends to be n-type at lower growth temperatures of approximately 850°C to 830°C, the carrier concentration being $n \simeq 2 \sim 0.5 \times$

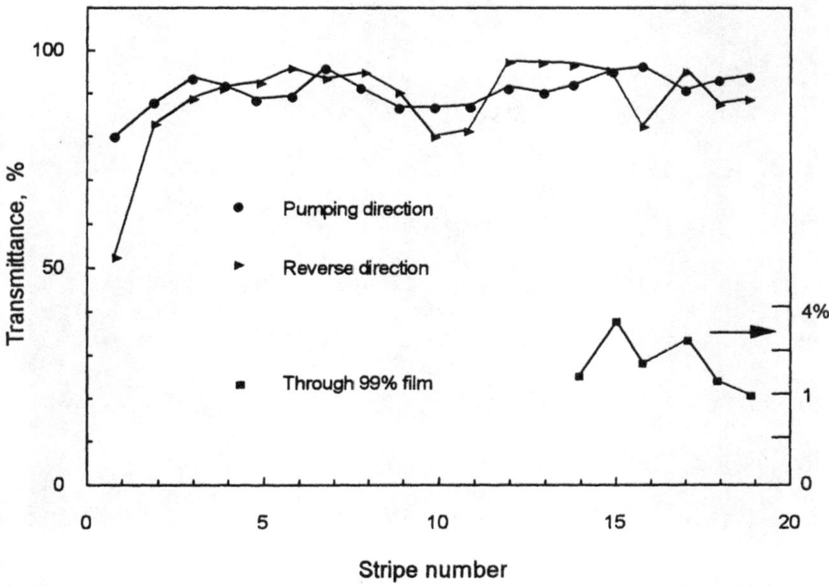

Figure 5.15 Waveguide transmittance at λ = 614 nm before (● and ▶) and after (■) formation of reflection films [7].

10^{16} cm^{-3}. We showed in Figure 3.20 that the absorption coefficient, α, of n-type epitaxial GaP is proportional to carrier concentration n, given by $\alpha/n \sim 0.05$ cm^{-1}/ 10^{16} cm^{-3} at the wavelength $\lambda = 1.1$ μm. Then, assuming that the absorption is due to free carriers for which α approximately varies as λ^2, the expected value of α at $\lambda = 630$ mm is almost 1/4 times the presently measured value. This discrepancy may be due to the difference in the conductivity type or the presence of deep levels that cause significant absorption at a shorter wavelength region.

5.2.2 Lasing Characteristics

The lasing experiment is essentially the same as that described in Chapter 3. An unfocused beam with a wavelength $\lambda = 1.064$ μm from a Q-switched YAG laser is incident on the window of the Raman laser at a small angle $\theta \simeq 1.7°$ relative to the resonator axis.

Figure 5.16 shows the lasing output power at the wavelength $\lambda = 1.112$ μm for various stripes with widths $s = 30 \sim 60$ μm formed on a single wafer as a

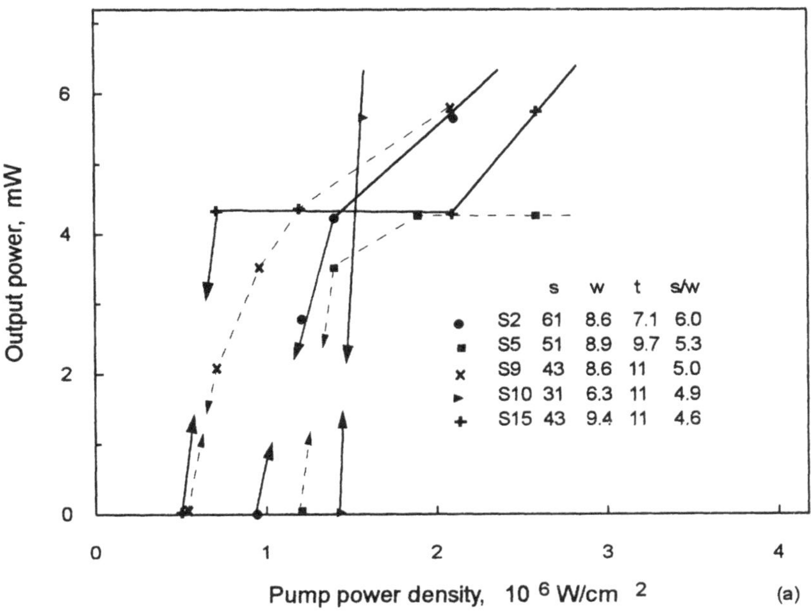

Figure 5.16 Output power of the Stokes radiation with wavelength $\lambda = 1.1$ μm as a function of (a) pump power density and (b) pump power, for several stripes in a wafer. stripe width s, window width w, and active layer thickness t are shown in (a) in units of μm [7].

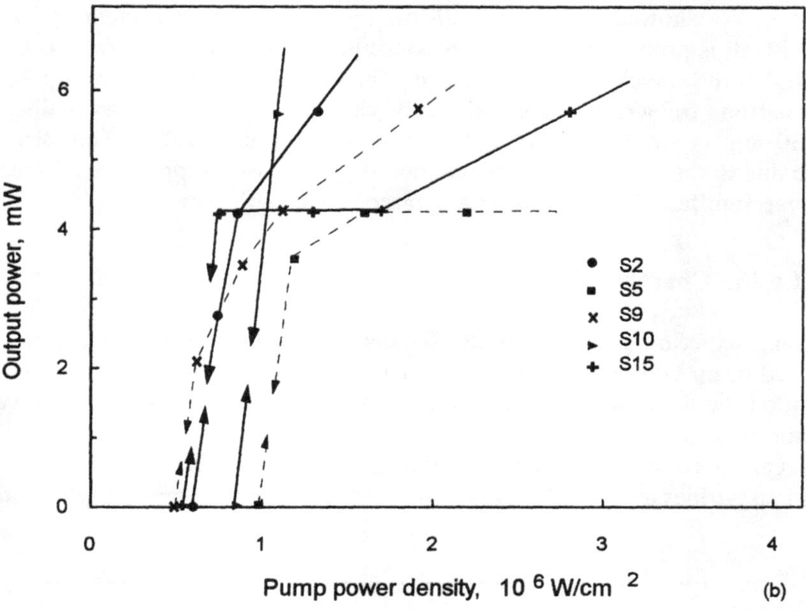

Figure 5.16 (Continued)

function of the incident pump power density and also as a function of the incident pump power. The lowest threshold pump power density is 0.5×10^6 W/cm^2, obtained for s = 40-μm stripes, which corresponds to a pump power of 500 mW. The above value of the threshold pump power density is comparable to or even less than that obtained for homogeneous GaP Raman lasers with large cross sections, 0.7×10^6 W/cm^2. Therefore, the optical loss at the hetero-interface is not significant in these stripes. On the other hand, the 30-μm stripes have shown higher threshold power densities, the best value being 1.4×10^6 W/cm^2, corresponding to a threshold power of 900 mW.

Here we must take into account the fact that the threshold pump power as well as pump power density depends on the ratio of the stripe width s to the window width w because the window will cause significant loss for the Stokes light. The threshold powers are replotted as a function of the s/w ratio in Figure 5.17, where it can be seen that the threshold powers for 30-μm stripes are on a single curve that rapidly rises at a smaller s/w ratio.

This tendency can be deduced from a simple calculation given below.

We use the subscripts s and i for the wavelength of the Stokes and pump radiations, respectively. The lasing condition is given by:

$$G^2 R_s^2 \, [(s - w)/s] \geq 1, \qquad (5.10)$$

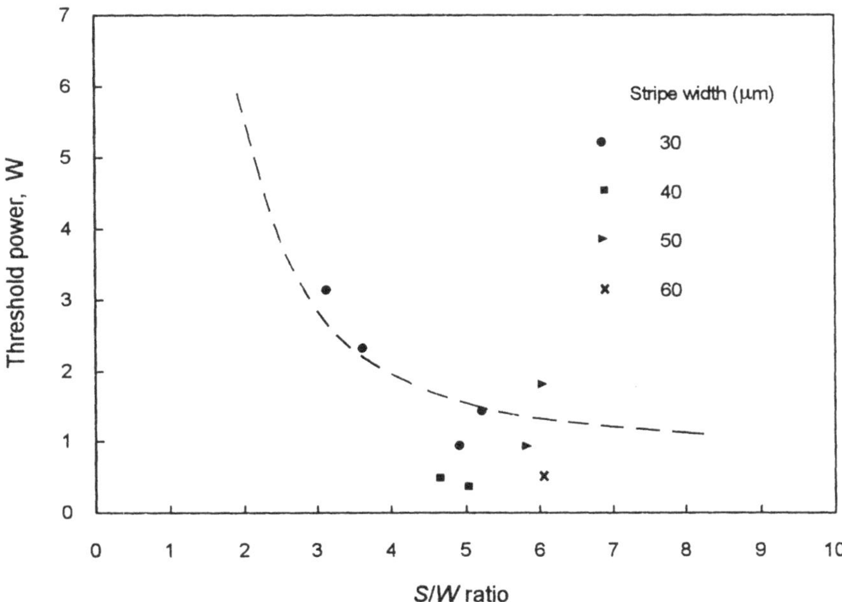

Figure 5.17 Threshold pump powers for various stripes as a function of s/w ratio. The dashed line is calculated for 30-μm stripes [7].

because the window loss at the input side is roughly given by $R_s(s - w)/s$ if we neglect the mode conversion loss for the fundamental mode of the Stokes field. The gain for a single pass is given by $G = e^{(g-a_s)l}$ and $g = Y|E_i|^2$, where E_i is the internal pump electric field. Then the following is obtained:

$$|E_i|^2 \geq \frac{1}{Y}\left(\frac{1}{2l}\ 1n \frac{1}{R_s^2} \frac{s}{s - w} + \alpha_s\right) \tag{5.11}$$

On the other hand, the input power $\boldsymbol{P}$ decreases after each round trip by a factor approximately given by:

$$r = e^{-2\alpha_i l} R_i^2 \frac{s - w}{s}. \tag{5.12}$$

Although only a finite number of reflections will contribute to the total pump power because of the pulse length, we simply assume infinite times of reflections and neglect the interference parts between reflected fields. Then, the average field is given by:

$$\frac{n}{2}|E_i|^2 = \frac{PT_i}{sd}\frac{1}{1-r}, \tag{5.13}$$

where n is the refractive index and T_i is the transmittance of the GaP surface.

The resulting threshold power $\boldsymbol{P}$ is given as follows:

$$P \geq \frac{n}{2}\frac{sd}{Y'T_i}\left(1 - e^{-2\alpha_i l}R_i^2\frac{s-w}{s}\right) \times \left(\frac{1}{2l}\ln\frac{1}{R_s^2}\frac{s}{s-w} + \alpha_s\right). \tag{5.14}$$

In this equation, s is the stripe width, w is the window width, d is the thickness of active layer, l is the length of laser, α is an effective absorption coefficient, R is the reflectivity of the input and output reflectors, T_i is the transmittance of the GaP surface, and Y' is a physical parameter that relates to Raman gain and also contains the loss from interfacial imperfections and other losses. The fitting parameter Y' is assumed here to have no relation to the stripe-to-window ratio. As shown in the dashed line in Figure 5.17, (5.14) can be fitted to the experimental threshold powers, with l = 3.7 mm, d = 10 μm, s = 30 μm, $\alpha_s = \alpha_i$ = 0.3 cm^{-1}, $R_s = R_i$ = 0.99, and Y' = 9.5 × 10^{-8} cmW^{-1}.

It can be seen that increasing the s/w ratio higher than the present values of approximately 5 to 6 is not effective for low-power operation. It is necessary to further reduce sidewall scattering by improving the interface smoothness, flatness, and angle to the epitaxial planes.

As a conclusion for the multimode pumped semiconductor Raman laser, opening of the window by the photolithographic technique limits the minimum size of the waveguide cross section, so further reduction of the threshold power will be a difficult problem. In the next section, we will describe a semiconductor Raman laser that needs no window opening. However, the multimode pumped semiconductor Raman laser has a great advantage in that we can widely change the pump light wavelength without changing the resonator films. Therefore, it will serve as a highly coherent light source pumped by lasers with a moderate power level of the order of 1W, such as a *cw* YAG laser or Ti-sapphire wavelength tunable laser.

5.3 RAMAN LASER PUMPED WITH A FUNDAMENTAL MODE

5.3.1 Structure

The most suitable pump source of the semiconductor Raman laser will be a laser diode because its frequency can be easily tuned over a few terahertz by simply changing the current. To use a laser diode as a pump source, however, the Raman laser should have a threshold pump power of less than of approximately 100 mW.

In this section, we describe a simple waveguide structure semiconductor Raman laser in which both the pump field and the Stokes field can propagate in a fundamental mode, thus ensuring the lowest loss propagation and eliminating the difficulty in the introduction of the pump beam [8, 9]. Low-threshold power operation on the 10-mW level can be expected for this single-mode pumped semiconductor Raman laser.

The structure is shown in Figure 5.18. The GaP core is made narrower than 10 μm, and the reflectivity of the film on the input side is made highly reflective at the wavelength of the Stokes light (1.11 μm) and highly transparent at the wavelength of the pump light (1.064 μm), as shown in Figure 5.18(b), so that there is no need to open a window. In this structure, the incident pump beam can propagate with a fundamental transverse mode and is reflected back at the exit face, which is highly reflective to the pump beam. On the other hand, in the structure described in Section 5.2, the pump beam essentially has a multi-transverse mode in the resonator. When a stripe becomes very narrow, the losses for the higher order modes are much larger than those of the fundamental mode, both for the pump field and the Stokes field. Also, the pump field can be resonantly enhanced when it is a single mode and the input resonator has a finite reflectance, as will be discussed later.

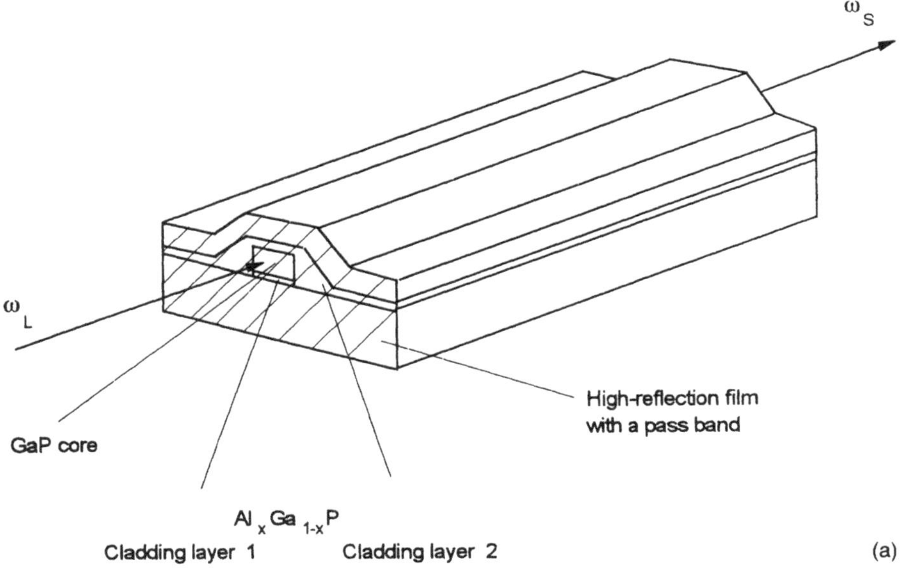

Figure 5.18 (a) Narrow stripe semiconductor Raman laser with reflection film having a pass band at pump wavelength. (b) Wavelength dependence of transmittance of reflection film on the input plane [8].

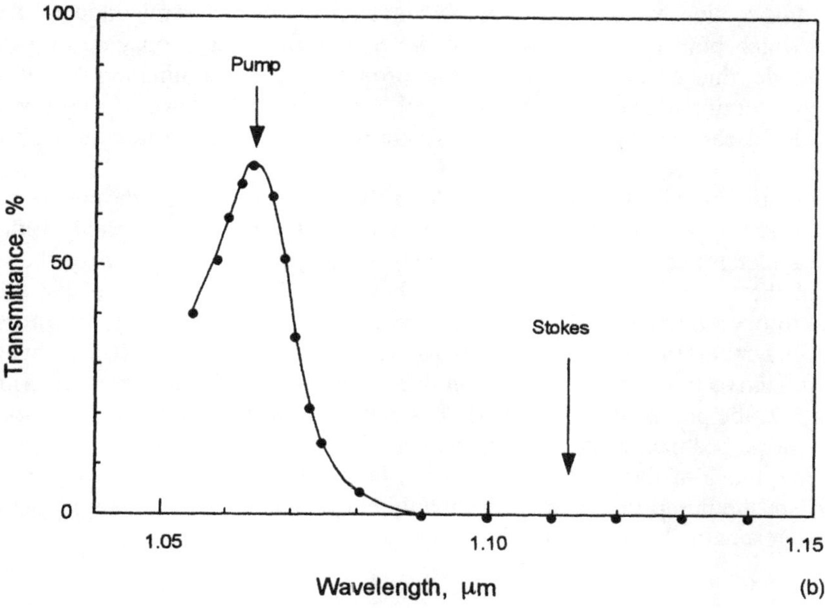

Figure 5.18 (Continued)

The reflection film at the input face is made of multilayers of SiO_2 and TiO_2 with each layer having a thickness of $\lambda/4$. The structure of the multilayers can be expressed as $GaP(L \cdot H)^m \cdot L \cdot L \cdot (H \cdot L)^m \cdot L$, where H and L means the high-refractive index layer (TiO_2) and the low-refractive index layer (SiO_2), and m is the number of times of repeating, which has been chosen to be 4.

When we need a finite reflectance at a pump wavelength, as will be later described, another configuration GaP $L \cdot (LH)^m \cdot L \cdot L \cdot (HL)^m$ is adopted. Figure 5.18(b) shows the transmittance, as a function of the wavelength, of the film deposited on a glass plate used as a monitor for the deposition process. It shows the characteristics of a narrow bandpass filter. However, its transmission band is about 5 THz, which is sufficiently broad for a pump source to be used in a heterodyne demodulator for a terahertz-modulated light signal.

The maximum transmittances at the pump wavelength range from 70% to 90% and are less than 1% at the wavelength of the Stokes light. On the other hand, the reflection film on the output side has transmittances of about 1% at both wavelengths.

The fabrication method for forming the GaP core and $Al_xGa_{1-x}P$ cladding layers is the same as described in Section 5.2, except that the width of the stripe is made narrower than 10 μm. The thickness of the core region ranges from 3 to 10 μm.

Figure 5.19(a) is an optical micrograph of a polished cross section of the stripe region; the crystal is illuminated uniformly from the back side. The figure shows excellent waveguiding characteristics of the GaP waveguide. The aluminum composition of the cladding layers is in the range $x \approx 0.1 \sim 0.2$. As was shown in Figure 5.3, this corresponds to $\sin \theta_c = 0.15 \sim 0.2$, where θ_c is the critical angle at the GaP—$Al_xGa_{1-x}P$ interface for the total reflection of a plane wave. In this range of $\sin \theta_c$, waveguiding occurs not only for the fundamental transverse mode but also for some higher order modes, although the loss would be much larger for higher order modes.

In the growth process for the waveguide shown in Figure 5.18(a), the stripes were formed by etching out whole regions except the stripes. However, the growth speed of the second cladding layer on the top surfaces of the stripes tends to be very slow. Sometimes, even the etching of the corners of the stripes takes place, while growth occurs only laterally at the sides of the stripes.

It is known that in an epitaxial growth process, some surface particles with a form of atoms or molecules migrate on an epitaxial surface until they find stable sites of the crystal lattice, which are called kinks or steps, and finally stick to these sites. Because the side regions of the stripes contain a high density of growth steps and kinks, growth proceeds mainly in these side regions in the lateral direction. Therefore, the growth rate is much higher in the lateral direction than in the vertical direction. This effect becomes pronounced when the stripe width is made less than 10 μm, which means that the migration distance of migrating species for $Al_xGa_{1-x}P$ growth on GaP is of this order of magnitude at the growth temperature 860 °C. When the sticking rate at the side regions is too high, the top surface region becomes unsaturated so that the etching of the stripes takes place. This causes a rugged surface and increases light-scattering loss of the Stokes light.

To overcome this difficulty, we have made an improvement in the stripe formation process; that is, only narrow recesses (width $= 10 \sim 20$ μm) at either side of each stripe are etched out. The formation of narrow recesses was first adopted in the fabrication of double-channel buried-heterostructure laser diodes [10].

Because the width of the recessed regions is made near to or narrower than the migration distance, these recesses are rapidly buried as a result of the lateral growth. Then, the growth on the stripe becomes easier, so the etching of the stripe regions is little observed. The cross-sectional growth patterns for the improved method are shown in Figure 5.19(b).

As another improvement, a slight wet etch (~ 60 nm) is carried out after reactive ion etching (RIE) has been made using an $H_2SO_4 : H_2O_2 : H_2O = 4 : 1 : 1$ solution instead of the previously used $HNO_3 : HCl : H_2O = 1 : 3 : 0$ to 4 solutions. It has been found that the latter solution greatly enhances the roughness of the corner regions of the stripes. Stripes with width as narrow as 6 μm are fabricated by this method, and they show excellent low-threshold characteristics, as is shown in Figure 5.23.

(a)

(b)

Figure 5.19 Cross-sectional views of waveguide semiconductor Raman lasers. Markers represent 10 μm; (a) broad recesses, (b) narrow recesses [9].

5.3.2 Lasing Characteristics and Design of the Resonator Films

Pumping is carried out using a *cw*-YAG laser, instead of the *Q*-switched YAG laser used for the bulk Raman laser, together with an optical modulator that produces optical pulses having a width ranging from 100 ns to 3 μs, as illustrated in Figure 5.20. The optical isolator inhibits a build-up of the incident YAG laser power resulting from the optical feedback effect. The YAG laser beam is finally focused on the cross section of the GaP core using a focusing lens that has a focal length l = 8 mm and a numerical aperture NA = 0.5.

Beam focusing is performed by observing the spot image on an infrared viewer screen using a microscope system composed of a beam focussing lens, beam splitter, and image formation lens. These components, except the beam focussing lens, can be removed away from the optical path during pumping experiment.

The minimum diameter of the beam spot at the input surface is approximately 4 μm. With proper positioning, the far field pattern of the transmitted pump power shows a fundamental transverse mode character. The lasing output beam is introduced to a monochromator and detected by using an InGaAs photodiode.

As will be clarified later, the pump field strength can be increased higher than that of the incident beam, so the effective threshold pump power is reduced if the input mirror of the resonator has a finite reflectance rather than 0% reflectance and the output mirror has nearly 100% reflectance. In the present experiment, the input mirror reflectance is chosen to be approximately 10% to 30%. The effect of the standing wave along the waveguide is observed when the temperature of the semiconductor Raman laser is changed by placing it on a Peltier temperature controlling

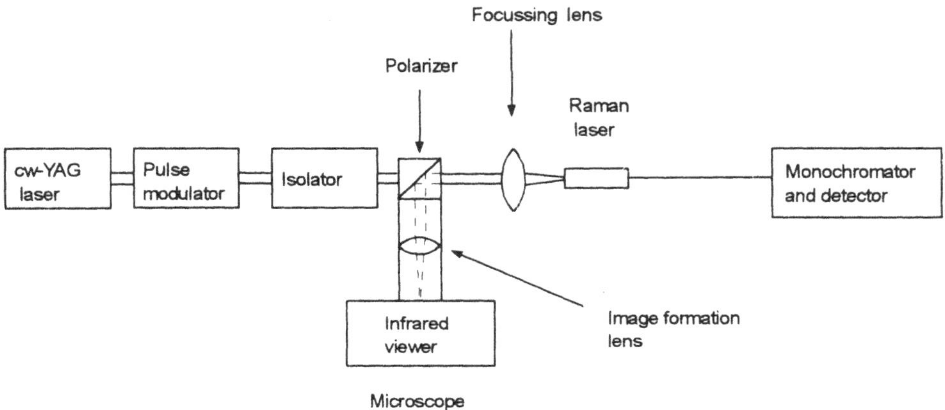

Figure 5.20 Experimental setup for lasing of a semiconductor Raman laser pumped by a pulse-modulated *cw*-YAG laser.

unit. Figure 5.21 shows the intensity of the pump power transmitting through the output plane having 99% reflectivity, as a function of temperature. As a result of the change in the refractive index of GaP, the peaks of the transmittance appear with a temperature spacing of 0.63 °C. The change in the index of refraction, Δn, corresponding to the temperature interval should be given by $\Delta n = \lambda_0/2\ell$, where ℓ is the length of the waveguide (4.2 mm) and λ_0 is the wavelength of the pump beam (1.064 μm). Hence, we obtain the temperature coefficient of the refractive index of the GaP core:

$$\beta = \frac{1}{n}\frac{dn}{dT} = 6.5 \times 10^{-5}\,\mathrm{K}^{-1} \quad \text{at} \quad \lambda = 1.064\ \mu m \tag{5.15}$$

This result agrees with the reported values of the temperature coefficient for a bulk GaP ($\beta \simeq 7.1 \times 10^{-5}\mathrm{K}^{-1}$ at $\lambda = 0.63\ \mu m$ and $\beta \simeq 5.6 \times 10^{-5}\mathrm{K}^{-1}$ at $\lambda = 1.115\ \mu m$) [8]. The intensity change, as shown in Figure 5.21, is seen clearly only when

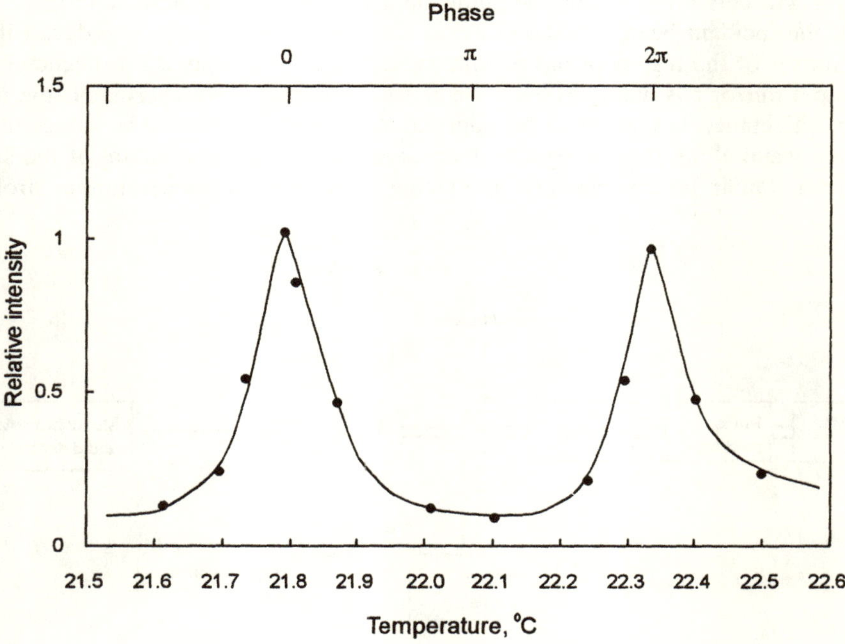

Figure 5.21 Transmittance of the pump beam through a GaP core with thickness 3.4 μm and width 9 μm, as a function of substrate temperature. The solid curve is calculated on an assumption that $r^2 = 0.25$ [9].

the far field pattern of the transmitted pump field shows a fundamental transverse mode.

The structure of the multilayers is expressed as GaP L · (L · H)^m · L · L · (H · L)^m. The first layer L on the surface of GaP has been introduced to obtain a desired finite reflectance ($R \leq 25\%$) at the pump wavelength $\lambda = 1.064$ μm.

We now address the significance of the finite reflectance value. When the fundamental transverse mode of the pump beam is transmitted through a waveguide having input and output field amplitude reflectances r and r', respectively, the intensity transmittance, T, through the waveguide is given by the following equation [9]:

$$T = \frac{(1 - r^2)(1 - r'^2)}{(1 + rr')^2} \frac{1}{1 + F \sin^2 \dfrac{\delta}{2}} \tag{5.16}$$

with

$$F = \frac{-4rr'}{(1 + rr')^2}$$

and

$$\delta = \frac{2\pi}{\lambda} 2nl$$

where F is the finesse, l is the length of the resonator, and n is the refractive index of the core. It is assumed that internal absorption and scattering losses can be neglected. In the present case, the intensity reflectance of the output surface is as high as $r'^2 \simeq 0.99$ (i.e., $r' \simeq -1$). Then, (5.16) becomes:

$$T \simeq (1 - r'^2)\left(\frac{1 + r}{1 - r}\right) \frac{1}{1 + F \sin^2 \dfrac{\delta}{2}} \qquad F \simeq \frac{4r}{(1 - r)^2} \tag{5.17}$$

Therefore, it can be seen that the internal pump power density increases by the factor:

$$k = (1 + r)/(1 - r), \qquad \text{at } \delta = 0, \tag{5.18}$$

provided r is low enough that the internal losses can be neglected. In the present experiment, the reflectance at the pump beam wavelength 1.064 μm has been chosen as $r^2 = 0.25$, (i.e., $r = 0.5$). The internal pump power is therefore expected to be three times greater than that for $r^2 = 0$; that is, $k \simeq 3$.

Figure 5.21 shows the pump light transmittance of a waveguide with width 9 μm and thickness 3.4 μm as a function of δ, which is measured by changing the substrate temperature. The measured finesse $F = 8.5$ corresponds to $r^2 = 0.26$, which is in agreement with the design value $r^2 = 0.25$. (At the Stokes wavelength = 1.119 μm, both input and output reflectances are $r^2 \simeq r'^2 \simeq 0.99$.)

Figures 5.22 and 5.23 show the lasing characteristics of various different stripes. In Figure 5.22, the stripes are formed with broad recesses, as shown in Figure 5.19(*a*); while in Figure 5.23, the stripes are formed with narrow recesses, as shown in Figure 5.19(*b*). In the case of the stripe with $s = 6$ μm and thickness $d = 3.4$ μm shown in Figure 5.23, the threshold optical input power is as low as 300 mW. Because the input spot diameter (4 μm) is slightly larger than the stripe thickness, the effective input power should be lower than 300 mW. On the other hand, the threshold pump power for the stripe with $s = 9$ μm is 600 mW, which is twice that of $s = 6$ μm. This result leads to the conclusion that the interface scattering loss at the etched sides is not yet dominant. Therefore, we will be able to further lower the threshold by

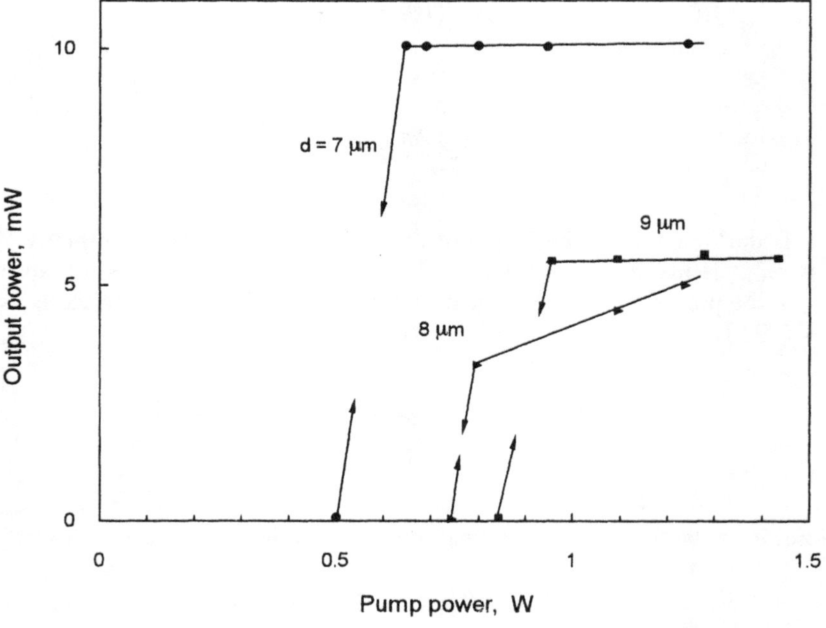

Figure 5.22 Lasing characteristics of three semiconductor Raman lasers with stripe width of 10 μm and stripe thicknesses of 7, 8 and 9 μm [8]. ● $t = 7$ μm; ▶ $t = 8$ μm; ■ $t = 9$ μm.

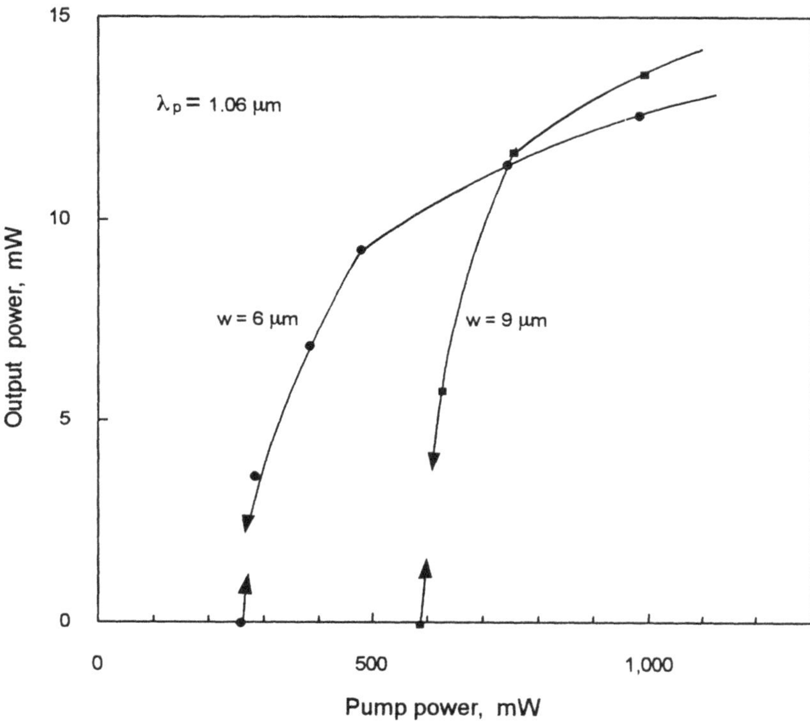

Figure 5.23 Lasing characteristics of two Raman lasers with thickness of 3.4 μm and stripe widths of 6 and 9 μm [9]. ● s = 6 μm; ■ s = 9 μm.

narrowing the stripe width if there is no limitation imposed by the spot size of the pump beam.

5.3.3 Estimation of Interface Loss for a Slab Waveguide

The most important loss is the internal absorption loss due to free carriers. It has been shown that for n-type GaP the absorption coefficient is proportional to the carrier concentration, given by $\alpha = 0.05\ (N/10^{16})\ \text{cm}^{-1}$ at $\lambda = 1\ \mu m$, where N is the free-carrier concentration per cm^3. TDM · CVP-grown GaP layers show n-type conductivity, with N ranging $2 \sim 0.5 \times 10^{16}\ \text{cm}^{-3}$ if oxygen contamination is carefully removed. This range corresponds to the absorption coefficient varying between $\alpha = 0.1 \sim 0.025\ \text{cm}^{-1}$.

The $Al_xGa_{1-x}P$ cladding layers are believed to contain more defects than GaP layers because $Al_xGa_{1-x}P$ growth has been made without vapor pressure control (see

Chapter 6). The conductivity tends to change from n-type to p-type, and the hole concentration tends to increase rapidly with increasing Al composition x. The x value at which the conductivity type changes depends on the growth temperature and tends to increase with increasing growth temperature. At $x \simeq 0.2$, the hole concentration frequently exceeds 2×10^{17} cm^{-3}. Therefore, the absorption loss will increase with decreasing thickness or width of the stripe, depending on the optical field confinement factor Γ.

We estimate how the absorption in the cladding layer affects the optical loss of the guided wave. We assume a slab waveguide with thickness d and infinite width. The electric field is described as:

$$E(x) = A \cos k_x x \qquad (-d/2 \leq x \leq d/2)$$

$$E(x) = B \exp\left\{-K\left(|x| - \frac{d}{2}\right)\right\} \qquad (x < -d/2 \text{ or } x > d/2) \tag{5.19}$$

The optical power confinement depends on the refractive-index difference n between the GaP core and the $Al_xGa_{1-x}P$ cladding layers. The refractive index of the $Al_xGa_{1-x}P$ depends on x, given by (5.1) at approximately $\lambda \simeq 1$ μm. The parameters in (5.19) can be determined by the same procedure as for the optical field in the heterostructure laser diode [12]. The power confinement factor Γ is defined as:

$$\Gamma = \int_{-d/2}^{d/2} |E(x)|^2 dx \bigg/ \int_{-\infty}^{\infty} |E(x)|^2 dx \tag{5.20}$$

Instead of Γ, we use the power confinement ratio given by:

$$\gamma = P_c/P_a = \left(\frac{1}{\Gamma} - 1\right) \tag{5.21}$$

where P_a and P_c are the optical powers in the Raman active layer and cladding layers, respectively. The effective absorption coefficient of the waveguide is given by:

$$\alpha_{eff} = \frac{P_a\alpha_a + P_c\alpha_c}{P_a + P_c} = \frac{\alpha_a + \gamma\alpha_c}{1 + \gamma} \tag{5.22}$$

where α_a and α_c are absorption coefficients in the active GaP layer and cladding $Al_xGa_{1-x}P$ layers, respectively.

Figure 5.24 shows γ and α_{eff} when it is assumed that the absorption coefficient of the cladding layers α_c is as high as 1 cm^{-1} (corresponding to $n = 2 \times 10^{17}$ cm^{-3}). It can be seen that α_{eff} starts to increase rapidly at approximately $d = 3$ μm when the absorption coefficient of the cladding layer is one order of magnitude higher than the absorption coefficient in the active layer. Therefore, in the case of Figure 5.23, with $d = 3.4\ \mu m$, the effect of free-carrier absorption in the cladding layers should not yet be prominent. However, the absorption loss in the cladding layer must be taken into account when we fabricate thinner active layers with $d < 3\ \mu m$.

The effect of the absorption in the cladding layers can be almost eliminated if the aluminum composition x is reduced to less than 0.1. However, the smallest thickness of the active layer for a given value of x is limited by the optical confinement factor of the pump beam; this means that thinning the active layer thickness to $\gamma \gtrsim 1$ is no use. For example, the limiting thickness for $x = 0.1$ is $d \approx 0.5\ \mu m$, as is seen in Figure 5.24.

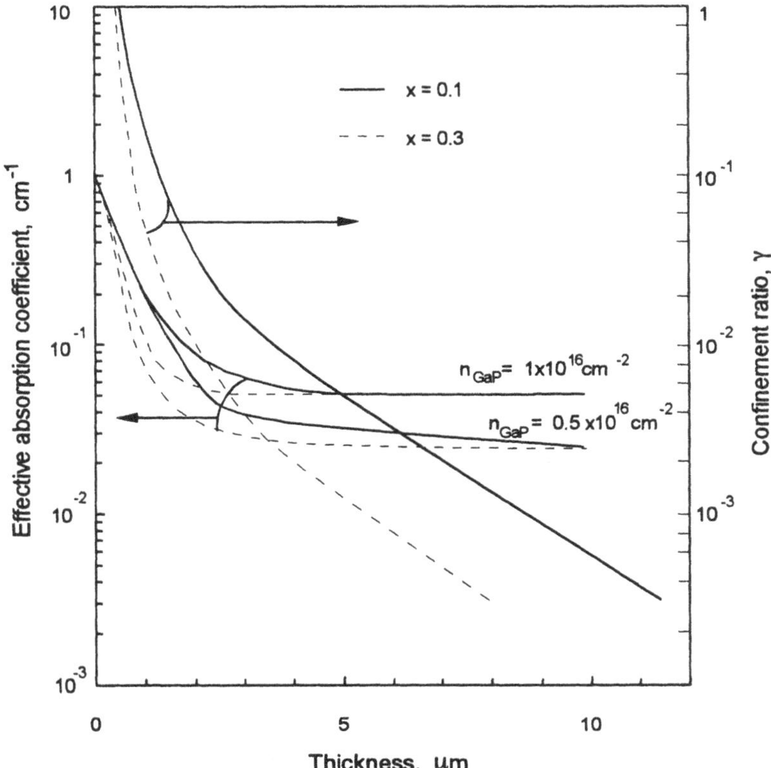

Figure 5.24 Calculated confinement factor γ and effective absorption coefficient α_{eff}. α_{eff}S for two different carrier concentrations of GaP cores, 1×10^{16} and 0.5×10^{16} cm^{-3}, are shown. In both cases, the absorption coefficient of the cladding layers is assumed to be 1 cm^{-1} [9]. ——— $x = 0.1$ --- $x = 0.3$.

In this discussion, the effect of the side walls has been neglected because the widths have been chosen to be much larger than the thickness. However, the side walls will generally contain a much higher number of defects because of damage produced by the RIE process. This effect will be discussed in Section 5.4.

5.4 RAMAN LASER AT SHORTER WAVELENGTHS

5.4.1 Effect of Additional Loss

For applications in optical communications, it is desirable to pump the Raman laser by a laser diode. In particular, the wavelengths of GaAlAs laser diodes for which high-power operation is available are in 800- to 900-nm region.

It is expected that the gain is proportional to the pump light frequency, while the wavelength dependence of the loss is more complicated. Although free-carrier absorption decreases at shorter wavelengths, we must expect additional losses, such as two-photon absorption or absorption via deep levels.

We describe the lasing performance of the semiconductor Raman laser with pump light wavelengths λ_p = 840 and 895 nm obtained from a titanium-sapphire laser [11].

It is found that there is an additional loss mechanism that has not been found for 1.064-μm pumping. The transmittance characteristics of the input resonator film are designed to have a maximum at a desired pump wavelength around 800 to 900 nm.

Figure 5.25 shows the setup for the lasing experiment. The *cw* titanium-sapphire laser beam is modulated to produce pulses with a width of 2.5 μs and a repetition frequency of 13 Hz, passes through an optical isolator, and is focused on the entrance surface of the waveguide. The diameter of the spot is 4 μm or less so that probably more than 90% of the beam power impinges on the entrance of the waveguide. The input surface has a reflectance of $R \simeq 0.3$, which causes the resonance effect. As a result, the internal pump intensity is expected to be higher than the input intensity (discussed in Section 5.3.2).

The transmitted pump light intensity shows oscillatory behavior as a function of the substrate temperature because of the finite reflectances $R_1 \simeq 0.3$ and $R_2 \simeq 0.99$ at the input and output planes, respectively. As a result, the temperature coefficient of the refractive index at $\lambda_p \simeq 840 \sim 890$ nm is found to be $\beta \equiv 1/n\ dn/dt \simeq 6.5 \times 10^{-5}\text{K}^{-1}$, which is nearly the same as that at λ_p = 1.064 μm.

However, it is found that the temperature rise due to absorption in the pumped waveguide cannot be neglected in this wavelength region. As shown in Figure 5.26, when the input power density is increased from 0.1 mW/μm^2 to the 1 mW/μm^2 level, the resonance frequency of the waveguide shows a shift corresponding to a substrate temperature shift of approximately 0.1 to 0.05 °C. In this transmittance

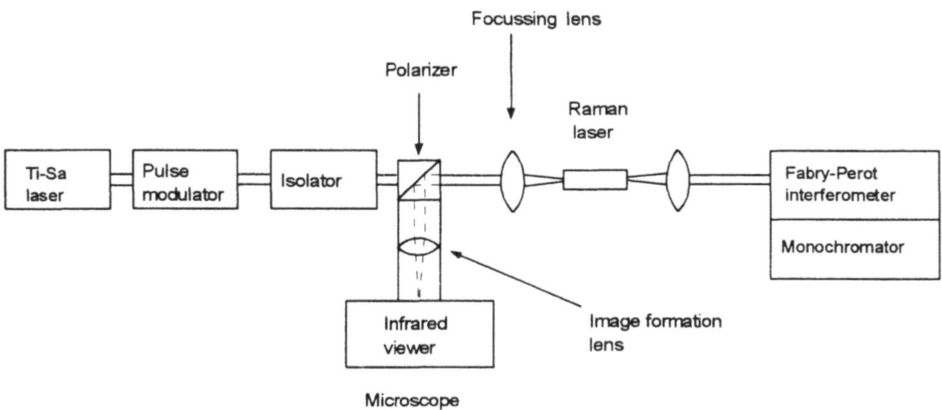

Figure 5.25 Schematic diagram of lasing experiment for a semiconductor Raman laser pumped by a ti-sapphire laser.

measurement, the input power is not pulse-modulated but in the *cw* mode. Although the temperature resolution of the Peltier controller is approximately 0.03 to 0.05C, it is found that the temperature shift for a 7-μm stripe is almost twice that of a 10-μm stripe.

Figure 5.27 shows the lasing characteristics for the Raman lasers with the pump wavelength λ_p = 840 nm and the output Stokes wavelength λ_s = 870 nm.

The Raman lasers are formed on a single wafer and have two different stripe widths, 10 μm (S24 and S22) and 7 μm (S23 and S21). For the 10-μm wide Raman lasers, the lasing characteristics are normal during a pump pulse width of 2.6 μs.

Figure 5.28 shows the spectral characteristics of the Stokes output power measured by a Fabry-Perot interferometer with a free spectral range of 15 GHz and half-width of 0.25 GHz. It is seen that they operate in a single longitudinal mode (longitudinal mode separation is 7 GHz). The observed spectral broadening of approximately 1 GHz is not the true linewidth but is due to a small temperature rise during the pump pulse duration of 2.5 μs. The temperature rise causes small shifts of the resonance frequencies for both the pump and Stokes waves.

On the other hand, all of the 7-μm wide Raman lasers have shown abnormal lasing characteristics. Lasing continues for only approximately 0.3 to 0.5 μs and stops after that during each pump pulse. At the same time, it is observed that the transmitted pump pulse intensity is reduced corresponding to the cessation of lasing. Therefore, lasing seems to stop because the pump light frequency is detuned from the resonance as a result of the temperature rise in the waveguide, so the internal pump field intensity decreases to below threshold. The threshold pump powers of most of the stripes with w = 7 μm are not smaller than those for the stripes with

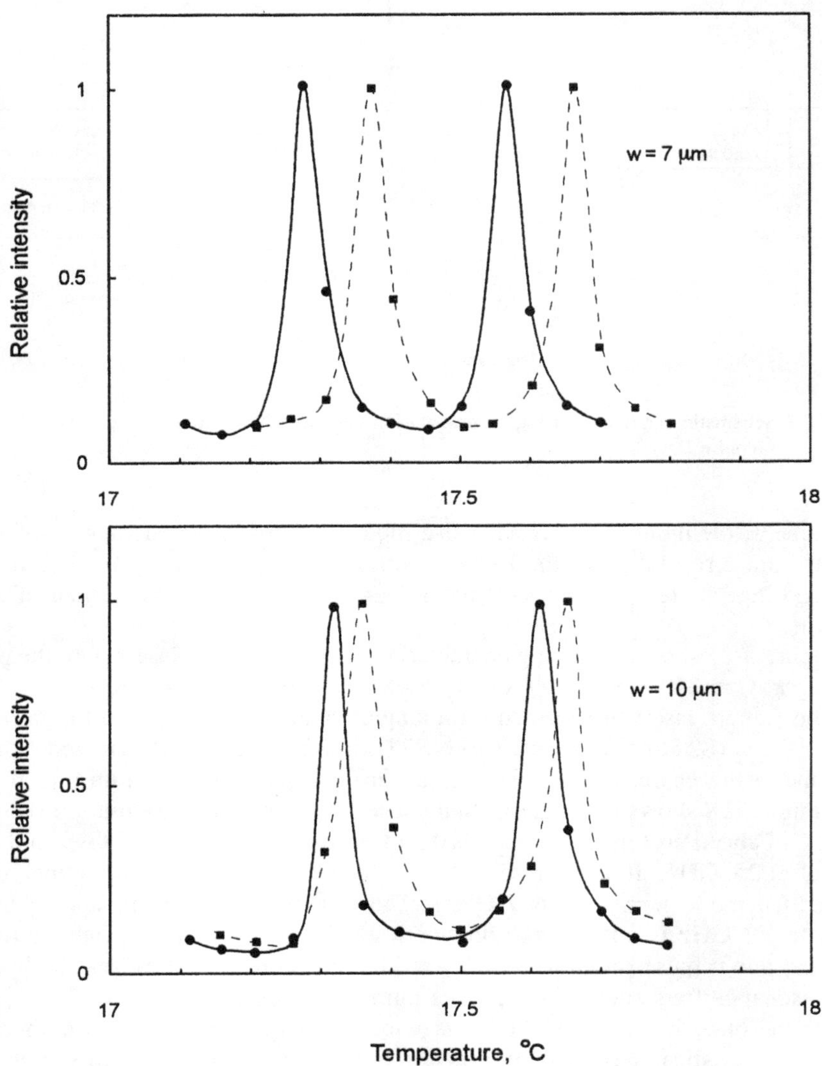

Figure 5.26 The transmitted pump light intensity as a function of the Peltier controller temperature for a 7-μm wide stripe (upper curves) and 10-μm wide stripe (lower curves). The solid and broken curves are for the input pump power densities 1 mW/μm^2 and 0.1 mW/μm^2, respectively [11].

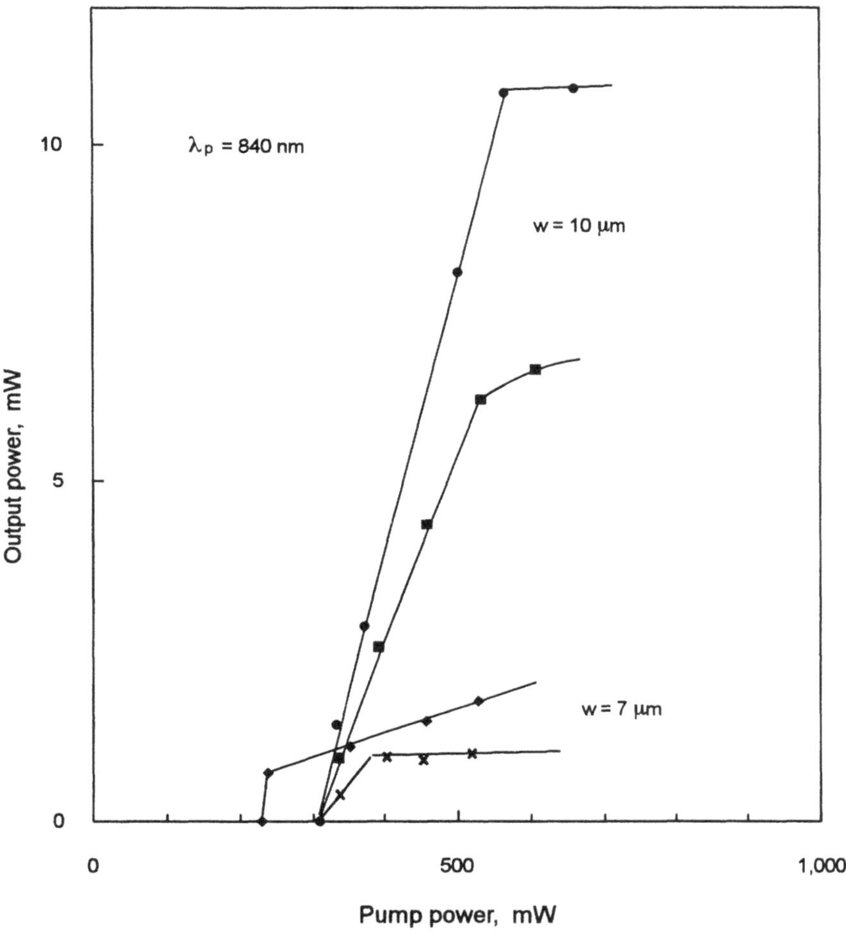

Figure 5.27 Lasing characteristics for λ_p = 840 nm and λ_s = 870 nm [11].

w = 10 μm. We do not observe such behavior when the pump wavelength λ_p = 1.064 m.

There are two possible absorption mechanisms that may occur in the shorter wavelength region: two-photon absorption and absorption via deep levels. The two-photon absorption edge at room temperature is thought to be 890 nm because the direct bandgap energy of GaP is E_g = 2.78 eV.

We have made a lasing experiment at λ_p = 895 nm (λ_s = 929 nm) to compare to the result for λ_p = 840 nm and thus determine the effect, if any, of two-photon

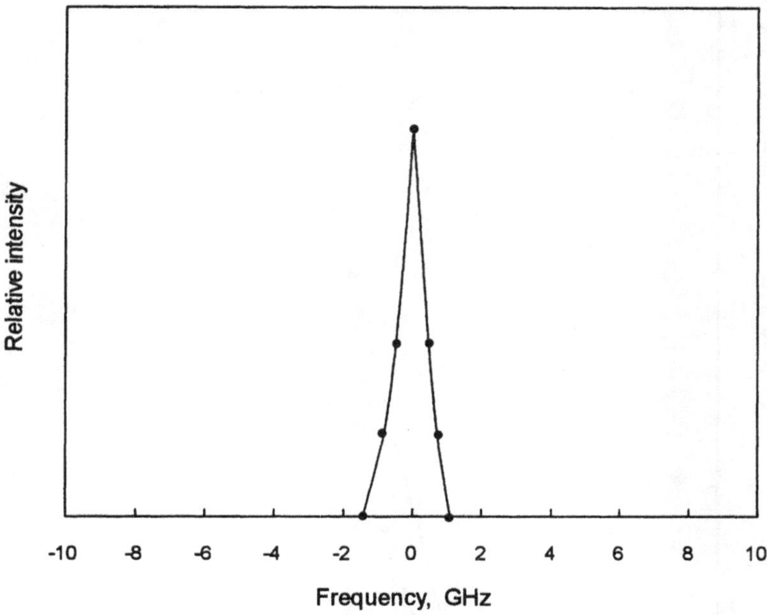

Figure 5.28 Output Stokes light transmission through a Fabry-Perot interferometer in pulsed operation [11].

absorption. Similar wafers containing stripes with w = 7 and 10 μm have been coated with resonator films having maximum transmittance at λ_p = 895 nm.

Figure 5.29 shows the result from a wafer that had shown low threshold performance. Lasing characteristics are quite similar to those for λ_p = 840 nm. For the stripes with w = 10 μm, the lasing is stable, and the threshold is as low as 300 mW. For the stripes with w = 7 μm, however, lasing stops after approximately 0.3 to 0.5 μs, and the threshold pump powers are always higher. From this result, we suppose that the temperature rise in the waveguide is not due to two-photon absorption but may be absorption via deep levels that have an absorption band at a wavelength region shorter than 1 μm. Because the absorption is strongly dependent on the width of the waveguide, it seems that the defects with deep levels have been produced through the RIE process. Although we have adopted a slight wet etching after the RIE process, a region with a high density of defects may yet remain.

Although the thresholds and the resonance frequency shifts are quite similar at the two pump wavelengths, the differential efficiency in the 7-μm waveguide at 840-nm pump wavelength is low compared to the result at 895 nm. There is a possibility that the two-photon absorption affects the efficiency of power conversion.

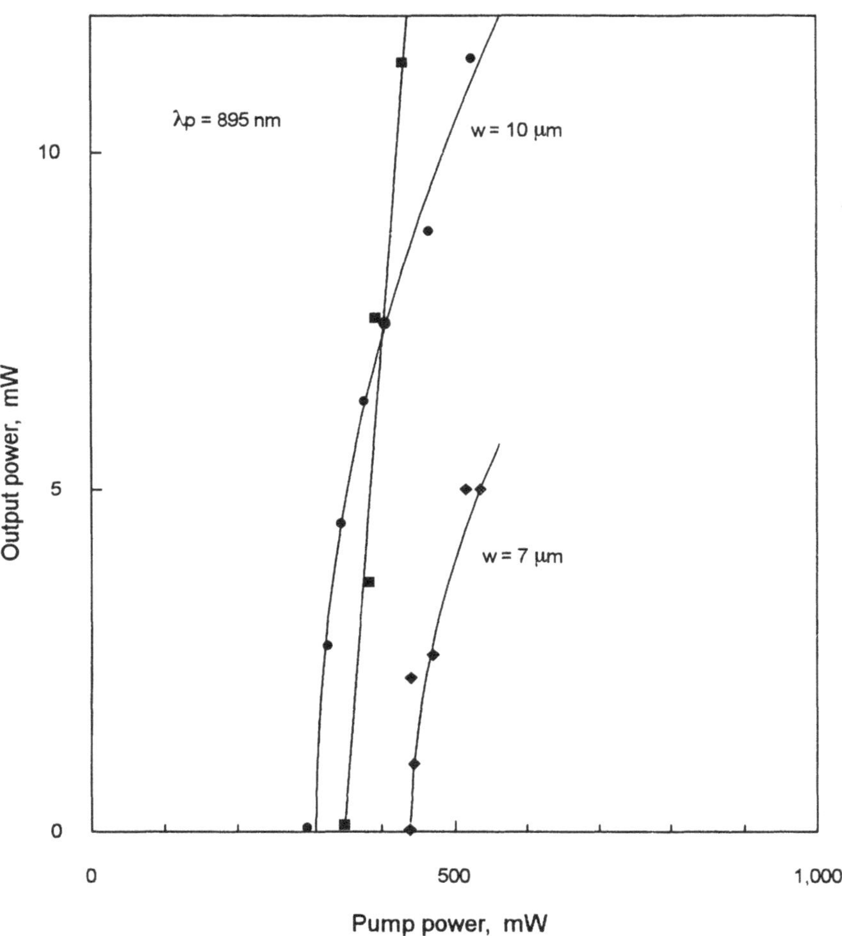

Figure 5.29 Lasing characteristics for λ_p = 895 nm and λ_s = 929 nm [11].

5.4.2 Estimation of the Sidewall Interface Loss

In Section 5.3.3, we discussed the relation between the optical confinement ratio P_a/P_c for a slab waveguide structure and the excess absorption at the interface $Al_xGa_{1-x}P$ region, where P_a and P_c mean the optical powers in the core and cladding regions, respectively. To estimate the effect of absorption at the side interfaces formed by the RIE process, we calculate the confinement ratio in the lateral direction. In the calculation, we assume that the refractive index of $Al_xGa_{1-x}P$ at approximately $\lambda =$ 900 nm is given by:

$$n_x = 3.143 - 0.352x \tag{5.23}$$

In comparison with the simplicity of the slab waveguide discussed in Section 5.3.3, exact analytic eigen solutions for the rectangular dielectric waveguide are not found. We simply assume that the optical field in a cross section of a waveguide is approximately given by $E = f(x) \cdot g(y)$, where $f(x)$ and $g(y)$ are field distributions in the thickness direction and lateral direction for a fundamental mode given by:

$$\begin{aligned}
f(x) &= B \exp\left\{-k\left(x - \frac{d}{2}\right)\right\} & \left(x > \frac{d}{2}\right) \\
f(x) &= A \cos k_x x & \left(\frac{d}{2} \geq x \geq -\frac{d}{2}\right) \\
f(x) &= B \exp\left\{k\left(x + \frac{d}{2}\right)\right\} & \left(x < -\frac{d}{2}\right) \\
g(y) &= D \exp\left\{-\gamma\left(y - \frac{w}{2}\right)\right\} & \left(y > -\frac{w}{2}\right) \\
g(y) &= C \cos k_y y & \left(\frac{w}{2} \geq y \geq -\frac{w}{2}\right) \\
g(y) &= D \exp\left\{\gamma\left(y + \frac{w}{2}\right)\right\} & \left(y < -\frac{w}{2}\right)
\end{aligned} \tag{5.24}$$

The initial defect concentration in the $A_xGa_{1-x}P$ will be higher than in the GaP because the growth is performed without any phosphorus vapor pressure control. Although the region with a high concentration of deep levels will extend into the sides of both the GaP and the $Al_xGa_{1-x}P$ in the lateral direction after the RIE process, we simply assume that the deep-level concentration is much higher in the $Al_xGa_{1-x}P$ side. The result of the calculation is given in Figure 5.30, which shows the effective absorption coefficient of the waveguide as a function of the waveguide width. The absorption coefficient in the interface region α_{int} is varied from 1 to 30 cm^{-1}, while the absorption coefficient in the GaP core is fixed at 0.05 cm^{-1}, which corresponds to the free-carrier concentration $N = 1 \times 10^{16}$ cm^{-3}, a typical value for the GaP layer.

It is found that the effective absorption coefficient of the waveguide rises at approximately $w \simeq 6 \sim 7$ μm when α_{int} exceeds approximately 10 to 30 cm^{-1}. It

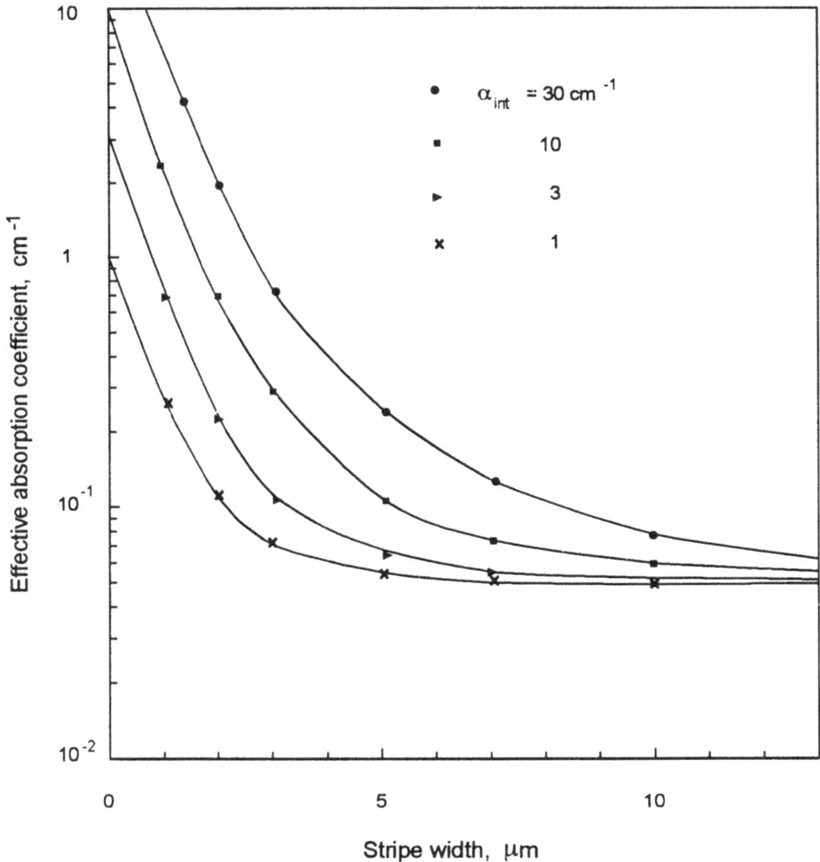

Figure 5.30 Calculated effective absorption coefficient of the waveguide as a function of the stripe width for different values of side interface absorption coefficient α_{int} [11].

should be noted that deep levels contributing to α_{int} are those within the thickness γ^{-1}, which is approximately 0.5 to 0.4 μm when $w \simeq 2 \sim 10$ μm. Although the nature of the deep levels is not known at present, deep levels having $\alpha > 10$ cm^{-1} are frequently observed in semiconductors. Wet etching processes are preferable for attaining a low deep-level concentration. However, it is difficult to make a vertical wall using only wet etching. The most suitable method will be a combination of the RIE process and a wet etching process.

5.5 TAPERED-WAVEGUIDE RAMAN LASER

Although it is possible to reduce the threshold pump power further by reducing the thickness and the width to less than those described in the earlier sections, the spot diameter of the pump beam must also be reduced. This will increase the requirements of the optics, such as the precision of the alignment. Also, the reflection loss of the Stokes light at the input and output planes will increase because the optical field broadens within the reflection films at the waveguide ends, which are as thick as approximately 1 to 2 μm. Tapered-waveguide structures will eliminate these difficulties. We proposed such a structure [13]. It should be noted that the tapered-waveguide laser diode was proposed and demonstrated [14, 15]. The tapered-waveguide Raman lasers are illustrated in Figure 5.31. In Figure 5.31(a), only the input region is tapered; whereas in Figure 5.31(b), both input and output regions are tapered. Both thickness and width can be tapered.

The tapered structure will cause an additional optical loss due to the mode conversion of the Stokes field from the fundamental mode to higher modes. A crude estimation of the loss can be made as follows.

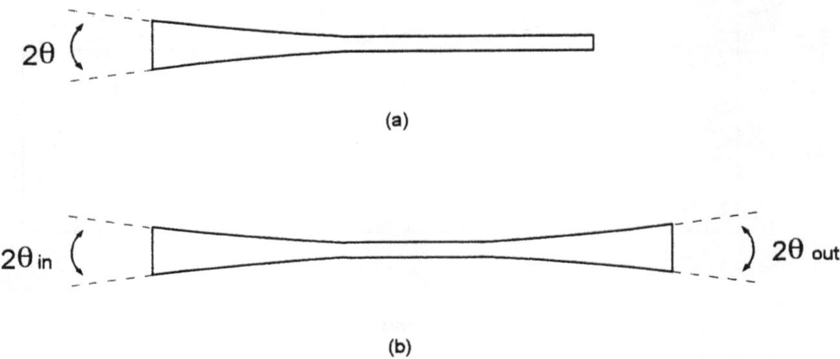

Figure 5.31 Schematic diagrams of tapered semiconductor Raman lasers [9].

We consider a waveguide having a symmetrical taper with a small angle 2θ in the thickness direction, as shown in Figure 5.31(a). The width of the core is assumed to be infinitely large. The mode conversion will occur each time reflection takes place at the end-faces, as well as the mode conversion along the waveguide.

We first estimate the mode conversion loss L_r for each reflection at the end-faces. The wave having a fundamental TE mode incident on the reflection plane is given approximately by:

$$E_{yi} = A \cos\left(\frac{\pi}{d} x\right) \exp(jk_{zo}z - j\omega t) \tag{5.25}$$

with

$$\left(\frac{\pi}{d}\right)^2 + k_{zo}^2 = k^2 = \left(\frac{2\pi n}{\lambda_0}\right)^2$$

as long as $d \gg \lambda_0/n$ holds. We have chosen the thickness direction to be x. The reflected wave E_{yr} can be expressed as a sum of the even modes, but we assume that only the second-order mode is generated. E_{yr} can be written as follows:

$$\begin{aligned} E_{yr} = A\Big\{&\sqrt{(1 - a_2^2)} \cos\left(\frac{\pi}{d} x\right) \exp\left(jk_{zo}z - j\omega t\right) \\ &+ a_2 \cos\left(\frac{3\pi}{d} x\right) \exp(jk_{z2} - j\omega t)\Big\} \end{aligned} \tag{5.26}$$

with

$$\left(\frac{3\pi}{d}\right)^2 + k_{z2}^2 = k^2$$

The reflection causes a change in momentum in the x direction Δk_x, which can be roughly estimated as:

$$\Delta k_x \simeq k \sin\theta = \frac{2\pi n}{\lambda_0} \sin\theta \tag{5.27}$$

The average momentum in the x direction for the field E_{yr} in (5.26) is given by:

$$\langle k_x \rangle_r = (1 - a_2^2)\left(\frac{\pi}{d}\right) + a_2^2\left(\frac{3\pi}{d}\right) \tag{5.28}$$

Therefore, the momentum change becomes:

$$\Delta k_x = a_2^2 \left(\frac{2\pi}{d}\right) \tag{5.29}$$

Equating (5.27) and (5.29), we obtain:

$$L_r = a_2^2 = \frac{nd}{\lambda_0} \sin\theta \approx \frac{nd}{\lambda_0}\,\theta \tag{5.30}$$

Next, the mode conversion loss L_l along the waveguide in the tapered region with length l can be estimated in a similar manner. The fundamental mode transmitting in the waveguide can be considered as being composed of two plane waves transmitting in zigzag passes, with angles $\pm\,\alpha$ where $\sin \alpha = (1/k)(\pi/d)$. Each reflection of a plane wave component at a boundary $x = d/2$ or—$d/2$ gives a change in the angle from α to $-(\alpha + \theta)$, instead of α to $-\alpha$. Therefore, each reflection gives the momentum change:

$$\Delta k_x = k \sin(\alpha + \theta) - k \sin \alpha \simeq k\theta \tag{5.31}$$

because α is assumed to be small. The above equation is the same as (5.29). This means that the mode conversion per reflection at a boundary $x = \pm d/2$ is the same as in (5.30).

On the other hand, the number of reflections of the plane waves at $x = +d/2$ in a length l of the waveguide is given by $N \approx l/(d/\theta)$, so the loss along a pass with length l is given by:

$$L_l \simeq N a_2^2 \simeq \frac{nl}{\lambda_0}\,\theta^2 \tag{5.32}$$

Therefore, the total loss due to the mode conversion by each one of tapers is given by:

$$L = (L_l + L_r) \simeq \left(\frac{nl}{\lambda_0}\,\theta^2 + \frac{nd}{\lambda_0}\,\theta\right) \tag{5.33}$$

where 2θ and l are the angle and length, respectively, of a tapered region.

We have fabricated waveguides with gradually tapered thicknesses, as shown in the inset of Figure 5.32. The growth of a tapered active layer was achieved by placing a GaP substrate in a furnace in such a way that temperature gradient, and hence the gradient of the growth rate, naturally followed the stripe direction, which

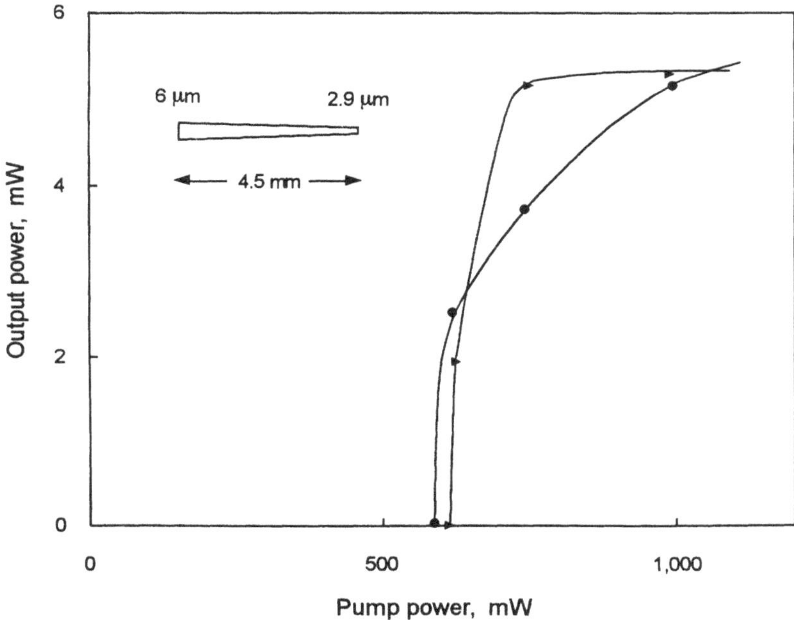

Figure 5.32 Lasing characteristics of two semiconductor Raman lasers with tapered waveguides, with thickness of 6.0 μm at the input end and 2.9 μm at the output end, and the stripe widths of 6 and 9 μm [9]. ● s = 6 μm; ▶ s = 9 μm.

is parallel to the direction of the axis of the furnace having temperature gradient. As a result, the thickness of the active layer was 6 μm at the input of the waveguide and 2.9 μm at the output, tapered over length of 4.5 mm.

To estimate the mode conversion loss, we use the values $\theta \simeq 0.33 \times 10^{-3}$, $l = 4.5$ mm, $d_i = 6\ \mu m$, and $d_0 = 3\ \mu m$ in (5.33), to obtain $L \simeq 6 \times 10^{-3}$. Therefore, it is estimated to be an order of magnitude smaller than the loss due to internal absorption and reflection at the end-faces.

The lasing characteristics for two stripes with widths s = 6 μm and 9 μm are shown in Figure 5.32. The threshold pump powers are probably 600 mW for both stripes, but we have yet to make a number of tapered-structure growths. There is, however, a possibility that the reflection loss of the Stokes wave at the output plane has increased because the thickness has become close to that of the resonator film. That is, the Stokes wave is inevitably broadened in the output film, resulting in the increase of the mirror-reflection loss for the Stokes wave.

It will therefore be necessary to introduce a structure like that shown in Figure 5.31(b), in which both the input and output regions are tapered to obtain remarkably low-power operation. In principle, both the thickness and the width of the active layer can be reduced to the limiting value determined from the optical confinement factor by introducing the tapered structure.

REFERENCES

[1] K. Suto, S. Ogasawara, T. Kimura and J. Nishizawa, "Semiconductor Raman Laser as a Tool for Wideband Optical Communications," *IEE PROC.*, Vol. 137, Pt. J, 1990, pp. 43–48.

[2] A.N. Pikhtin and A.D. Yaskov, "Refraction of Light in Semiconductors (Review)," *Sov. Phys. Semicond.*, Vol. 22, 1988, pp. 613–626.

[3] K. Suto, T. Kimura and J. Nishizawa, "Heterostructure Semiconductor Raman Laser," *IEE PROC.*, Vol. 134, Pt. J, 1987, pp. 215–220.

[4] K. Suto, T. Kimura and J. Nishizawa, "Heteroepitaxy of GaP-$Al_xGa_{1-x}P$ System by the Temperature Difference Method Under Controlled Vapor Pressure (TDM-CVP)," *J. of Crystal Growth*, Vol. 99, 1990, pp. 297–301.

[5] M. Bettini and M. Cardona, "Spatial-Dispersion-Induced Birefringence in Cubic Semiconductors," *Proc. 11th Int. Conf. Physics of Semiconductors*, Warsaw, Poland, 1972, pp. 1072–1077.

[6] K. Suto, T. Kimura and J. Nishizawa, "Lateral Optical Confinement of the Heterostructure Semiconductor Raman Laser," *Appl. Phys. Lett.* Vol. 51, 1987, pp. 1457–1458.

[7] K. Suto, S. Ogasawara, T. Kimura and J. Nishizawa, "Buried-Heterostructure Semiconductor Raman Laser With Threshold Pump Power Less Than 1W," *J. Appl. Phys.*, Vol. 66, 1989, pp. 5151–5155.

[8] K. Suto, T. Kimura and J. Nishizawa, "Semiconductor Raman Laser With Resonator Film Transparent to Pump Light," *IEE PROC.*, Vol. 138, Pt. J, 1991, pp. 396–400.

[9] K. Suto, T. Kimura and J. Nishizawa, "Semiconductor Raman Laser Pumped With a Fundamental Mode," *IEE PROC.*, Vol. 139, Pt. J, 1992, pp. 407–412.

[10] I. Mito, M. Kitamura, K. Kobayashi, S. Murata, M. Seki, Y. Odagiri, H. Nishimoto, M. Yamaguchi and K. Kobayashi, "InGaAsP Double-Channel-Planar-Buried Heterostructure Laser Diode (DC-PBHLD) With Effective Current Confinement," *IEEE J. Lightwave Tech.*, Vol. LT1, 1983, pp. 195–202.

[11] K. Suto, T. Kimura and J. Nishizawa, "Semiconductor Raman Laser With Pump Light Wavelength in the 800 nm Region," *J. Electrochem. Soc.*, Vol. 140, 1993, pp. 1805–1808.

[12] H.C. Casey, Jr. and M. B. Panish, *Heterostructure Lasers*, New York: Academic Press, 1978.

[13] J. Nishizawa and K. Suto, "Semiconductor Raman Laser," Japanese Patent 1793337, Application 1987.

[14] T.L. Koch, U. Koren, G. Eisenstein, M.G. Young, M. Oren, C.R. Giles and B.I. Miller, "Tapered Waveguide InGaAs/InGaAsP Multiple-Quantum-Well Lasers," *IEEE Photon. Technol. Lett.*, Vol. 2, 1990, pp. 88–90.

[15] G. Bendelli, K. Komori and S. Arai, "Gain Saturation and Propagation Characteristics of Index-Guided Tapered-Waveguide Travelling-Wave Semiconductor Laser Amplifiers (TTW-SLA's)," *IEEE J. Quant. Electron.*, Vol. 28, 1992, pp. 447–458.

Chapter 6
Fabrication Method of the Semiconductor Waveguide

6.1 PRINCIPLE OF STOICHIOMETRY-CONTROLLED LIQUID PHASE EPITAXY

To achieve low-threshold heteroepitaxial Raman lasers, as well as low-loss semiconductor waveguides for light modulators, we need to grow extremely high quality layers of GaP—$Al_xGa_{1-x}P$. Small optical losses like free-carrier absorption or deep-level absorption can seriously increase the threshold pump power because the order of Raman gain is only 0.1 cm^{-1}. Free-carrier concentration must be less than 2×10^{16} cm^{-3}. Also, concentration of deep levels arising from nonstoichiometric point defects like phosphorus vacancies or interstitial atoms should be reduced to a minimum. Although there may be a few different methods for heteroepitaxial growth of GaP—$Al_xGa_{1-x}P$ layers, a liquid phase epitaxy called the temperature difference method under controlled vapor pressure (TDM · CVP), which was found by Nishizawa, is the most suitable growth method [1–5]. It enables stoichiometry-controlled crystal growth by application of vapor pressure of a volatile element over the solution.

Figure 6.1(a) is a schematic diagram to explain the growth principle, while Figure 6.1(b) illustrates a practical liquid phase growth apparatus for single layer epitaxy. The growth well contains Ga solution and GaP poly crystals. Temperature difference is established in the melt by using a subsidiary heater. Poly crystals are continuously dissolved at an upper portion of the solution maintained at a higher temperature, and dissolved particles are transported on the surface of a substrate crystal by diffusion caused by the temperature difference. Therefore, there is no need to have a slow cooling process for segregation as in the case of conventional liquid phase epitaxy.

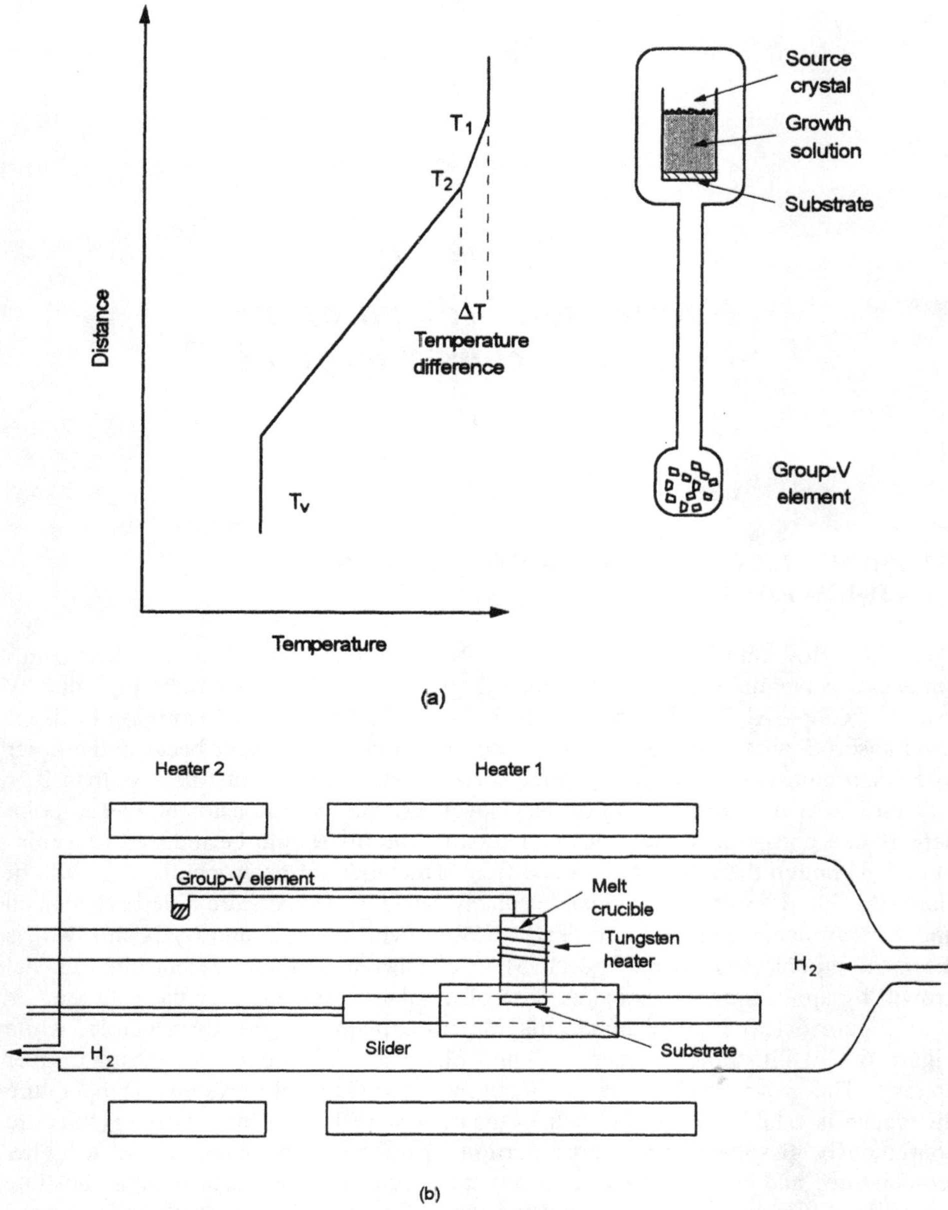

Figure 6.1 (a) Schematic diagram for the temperature difference method (TDM) under controlled vapor pressure (CVP). (b) Apparatus for liquid phase epitaxy by the TDM-CVP method [1,2].

Figure 6.2 shows an example of the relation between the growth speed and the subsidiary heater power, which causes the temperature difference in the solution. An epitaxial wafer surface contains very flat, table-like regions called facets where the growth rates are higher than those at the surrounding regions. Figure 6.2 shows the thicknesses at the facet regions and the growth thicknesses averaged over the epitaxial plane. Facet growth occurs as a result of two-dimensional growth mechanism, which is most typically observed in vapor phase epitaxy [6].

On the other hand, there is a chamber containing a group V element (phosphorus) connected to the solution well with a thin quartz tube, which supplies a vapor pressure on the top surface of the solution. The applied vapor pressure P_a is given by:

$$P_a = P_0 \sqrt{T_1/T_V} \tag{6.1}$$

where P_0 is the vapor pressure of the group V element. The vapor pressure of the phosphorus element (usually of P_4 molecules) is given in [7].

Extremely high quality crystals with exact stoichiometry can be grown by applying an optimum vapor pressure.

Figure 6.3 shows the cleavage planes of the layers grown at various vapor pressures [1,2].

Growth temperature was intentionally changed between 820° and 840°C during growth, as illustrated in the inset, in order to check the stability of the growth. The cleaved surface of the grown layer is very smooth at an optimum vapor pressure P = 100 Torr, while under other vapor pressures, 1, 10, and, 1,000 Torr, root-like faults are clearly observed. These faults may be defect-supersaturated regions. Actually, the dislocation density shows a sharp minimum at the same vapor pressure, which we call the optimum vapor pressure, as shown in Figure 6.4 [3].

Deep-level concentration also shows a minimum at the optimum vapor pressure as shown in Figure 6.5 [8,9].

The optimum vapor pressure that gives a high-quality grown layer was found to be a function of growth temperature as shown in Figure 6.6, which can be expressed as follows [1–3]:

$$P_{GaP} = 4.6 \times 10^6 \exp\left(-\frac{1.05eV}{kT_g}\right) \text{Torr} \tag{6.2}$$

Stoichiometry control by vapor pressure application was investigated for GaAs in more detail [2,3]. For GaAs, the optimum arsenic vapor pressure was found to be:

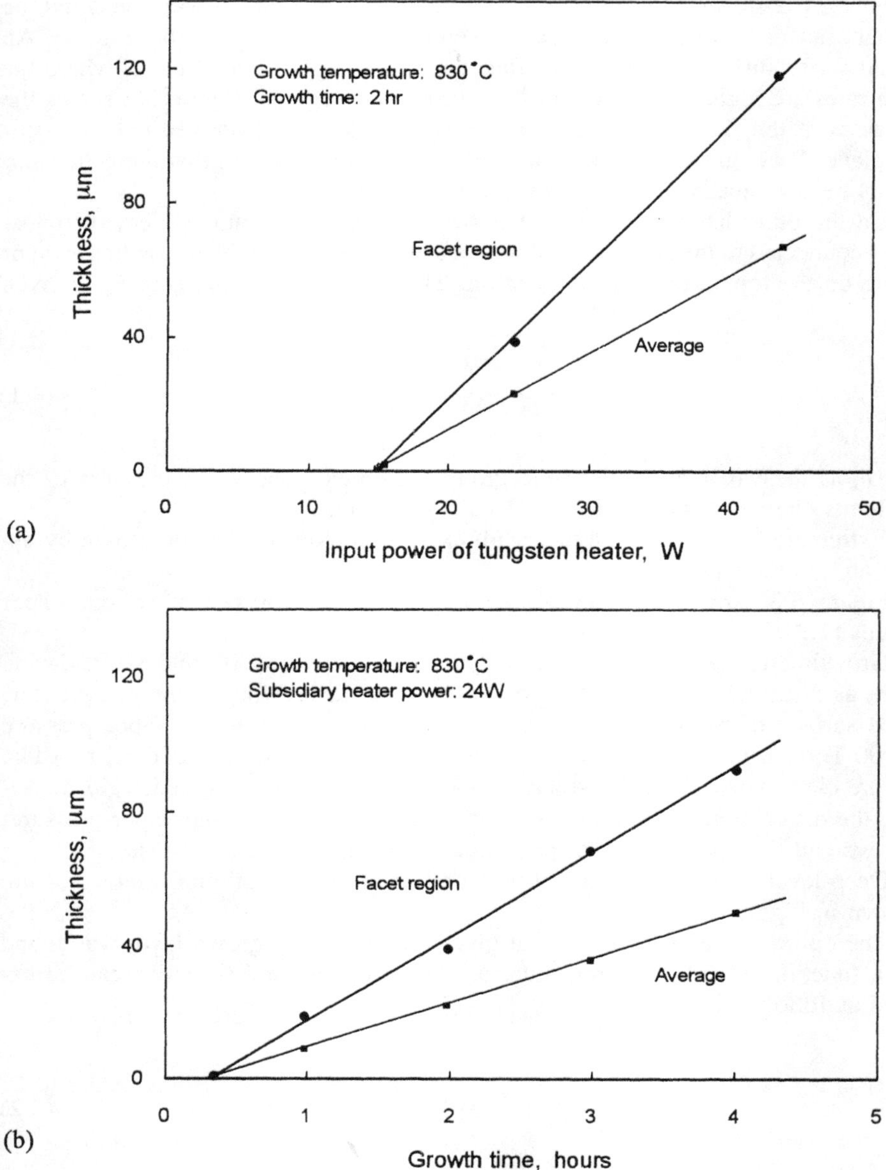

Figure 6.2 (a) Relation between tungsten heater power and growth thickness. (b) Relation between growth time and growth thickness. Tungsten heater power of 24W produces temperature difference of about 20 deg in melt [1].

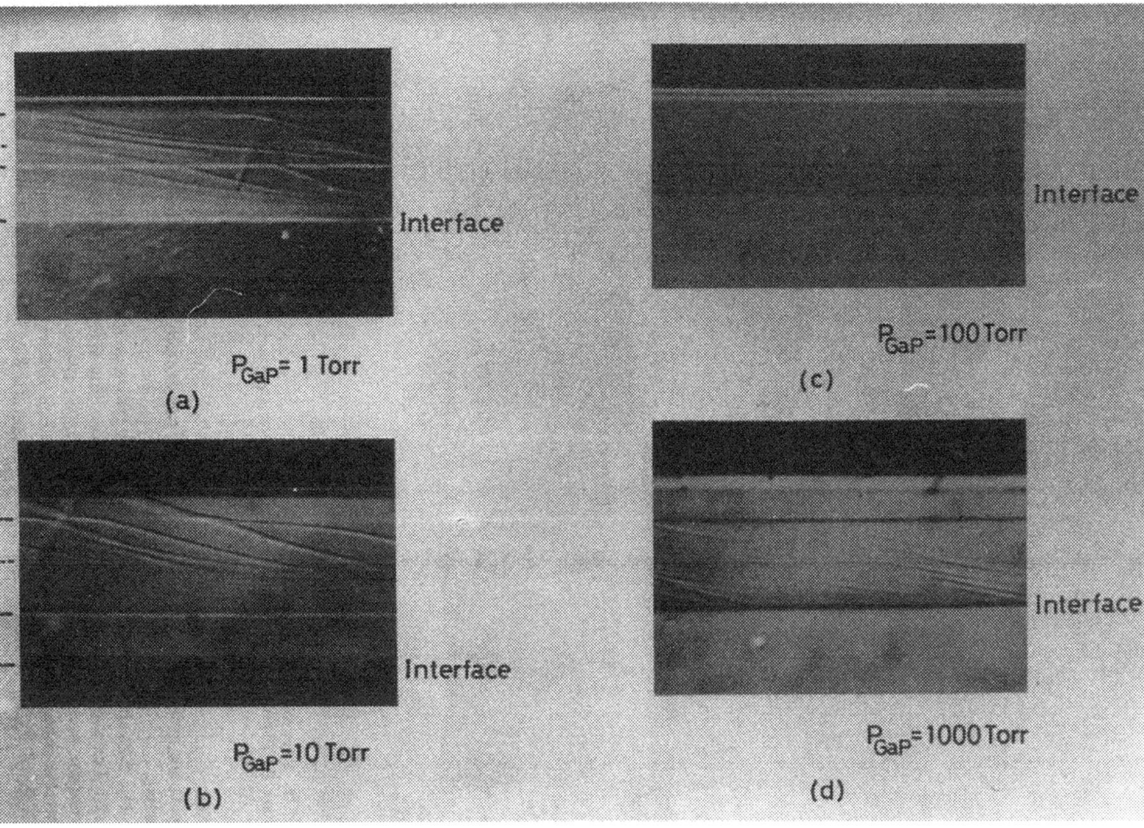

Figure 6.3 Cleaved surfaces of epitaxial layers grown at phosphorus vapor pressure of (a) 1 torr, (b) 10 torr, (c) 100 torr, and (d) 1,000 torr [1,2].

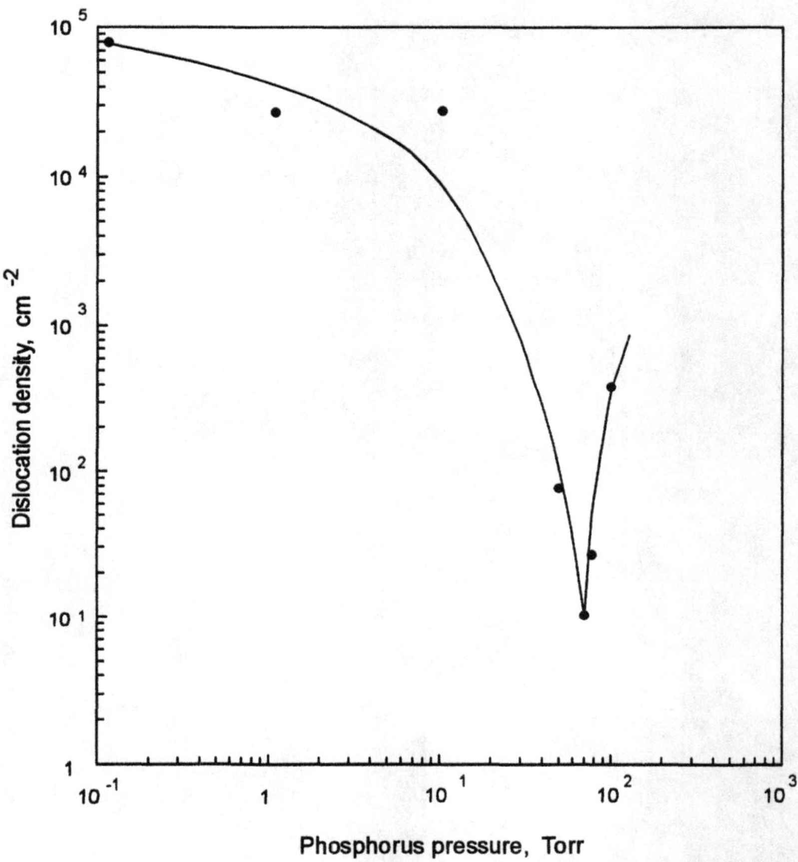

Figure 6.4 Phosphorus vapor pressure dependence of dislocation density in GaP epitaxial layer. A nearly dislocation-free layer is obtained at P = 67 torr when the growth temperature T_g = 800°C. The dislocation density in the substrate is on the order of $10^5/cm^2$ [3].

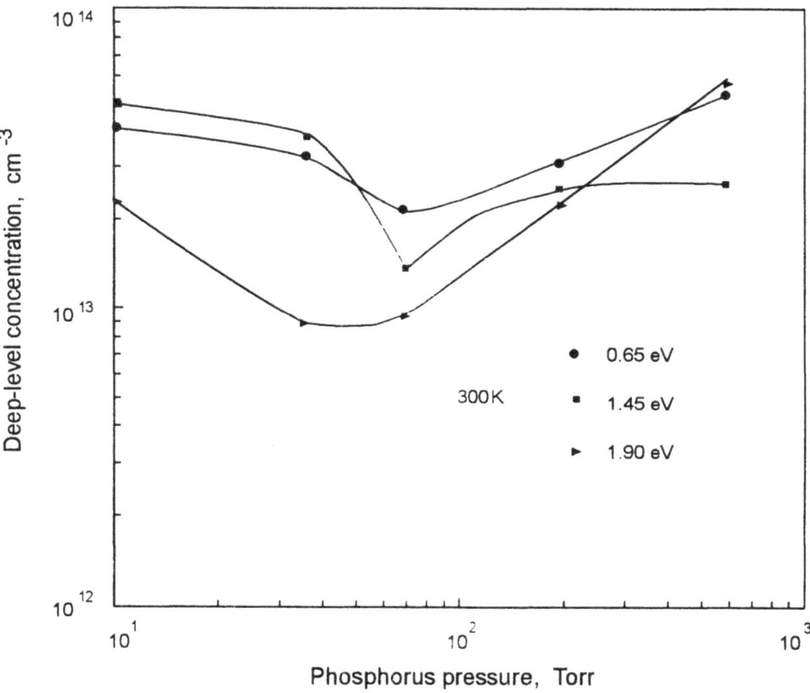

Figure 6.5 Relation between phosphorus pressure and the concentrations of deep levels at 0.65, 1.45, and 1.90 eV observed at 300 K [9].

$$P_{GaAs} = 2.6 \times 10^6 \exp\left(-\frac{1.01eV}{kT_g}\right)\text{Torr} \tag{6.3}$$

It was shown that the vapor pressure gives an equality of the chemical potential of the group V element between the gas phase, liquid phase, and solid phase [3,4]:

$$\mu_V^g = \mu_V^l = \mu_V^s \tag{6.4}$$

In other words, the saturation solubility of the liquid is no longer a constant at a given growth temperature even for a binary system, but it slightly increases until the three-phase equilibrium given in (6.4) is reached under an applied vapor pressure.

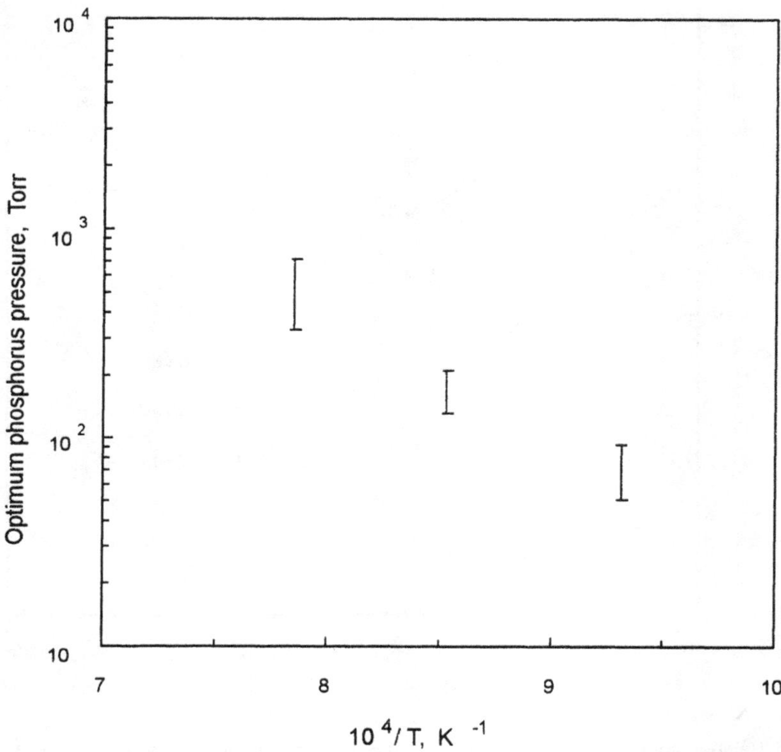

Figure 6.6 Relation between optimum phosphorus pressure and crystal growth temperature [1].

6.2 $Al_xGa_{1-x}P$—GaP HETEROEPITAXY

6.2.1 Growth Procedure

For the liquid phase epitaxy of $Al_xGa_{1-x}P$ layers, we need the ternary phase diagram of the Al—Ga—P system at a relatively low temperature below 1000°C. Tanaka, Sugiura, and Sukegawa investigated the Al—Ga—P phase diagram in the temperature range of approximately 800° to 1040°C. Single heteroepitaxial layers of $Al_xGa_{1-x}P$ were grown on GaP (111)B substrates by the temperature difference method [10].

Liquidus and solidus data by Tanaka and others and Ilegems and Panish are shown in Figure 6.7. The figure shows that the segregation coefficient of Al is very large, similar to the case of the Al—Ga—As system. In practical liquid phase epitaxy of $Al_xGa_{1-x}P$ layers, however, we are not exactly based on the phase diagram data, which are valid when the equilibrium is established in the whole system.

To eliminate occurrence of poorer morphology of a grown layer due to aluminum oxidation, aluminum metal is added just before the first run but not at a preceding melt-alloying stage. As a result, the aluminum composition of the grown layer becomes considerably smaller than that simply predicted from the phase diagram data because aluminum is consumed by forming a solid phase at the source region. Figure 6.8 shows the aluminum composition, x, of the grown layers for the first growth run. Each grown layer shows no detectable change in x, both in thickness and lateral directions.

However, a waste of aluminum as a result of alloying with source GaP causes a gradual decrease of x with increasing number of growth runs.

The growth of buried heterostructure for Raman lasers is achieved by the following processes: The first $Al_xGa_{1-x}P$ cladding layer and a GaP waveguiding layer are grown successively on a GaP substrate with a (100) surface. Then, the layers

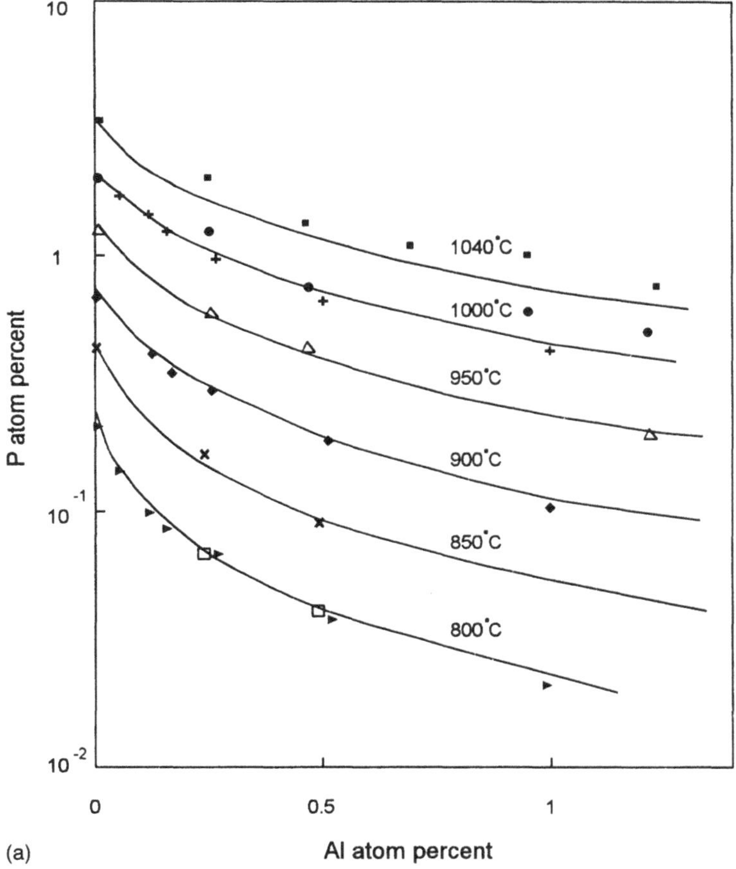

Figure 6.7 Isotherms in the Ga-Al-P system: (a) liquidus and (b) solidus [10].

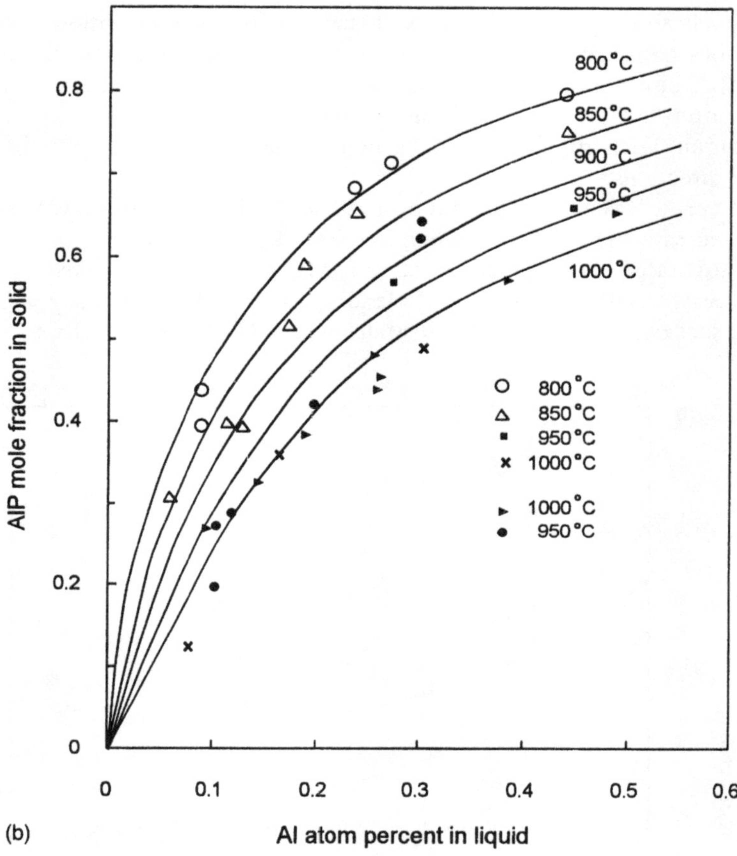

Figure 6.7 (Continued)

are etched away by reactive-ion etching using PCl_3 gas, leaving stripes along a ⟨100⟩ direction with widths of typically 10 mm and with an interval of 300 mm. Finally, the second $Al_xGa_{1-x}P$ cladding layer and a GaP overlayer are grown successively. We have shown the cross section of a buried heterostructure in Figure 5.19.

Figure 6.9 illustrates the growth apparatus for successive epitaxy of $A_xGa_{1-x}P$ and GaP layers on a GaP substrate. The optimum phosphorus vapor pressure is not known for $Al_xGa_{1-x}P$ crystal so that the phosphorus vapor pressure is applied only for the GaP growth.

Growth speed generally decreases with increasing x_{Al} of the grown layer, but other factors, such as the location of growth wells in the furnace, more strongly affect the growth speed.

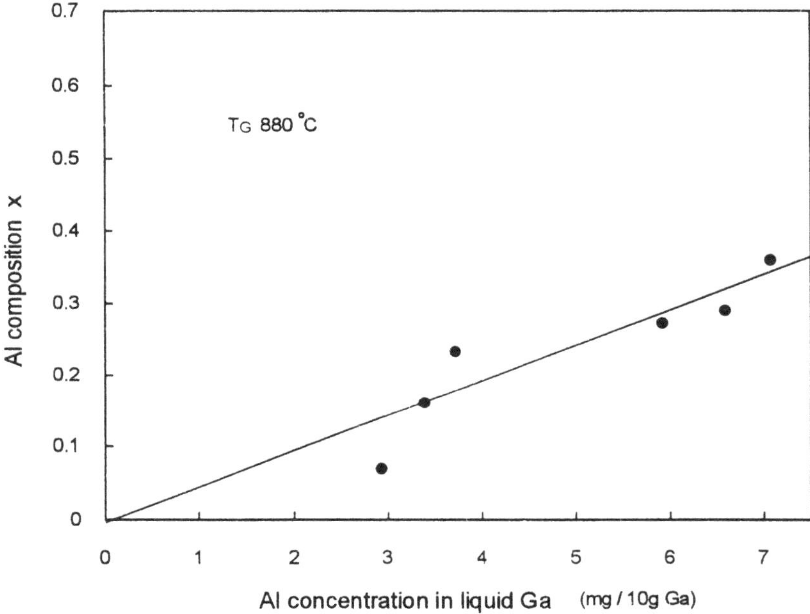

Figure 6.8 Relation between the aluminum composition, *x*, of a growth layer and aluminum concentration in liquid [12].

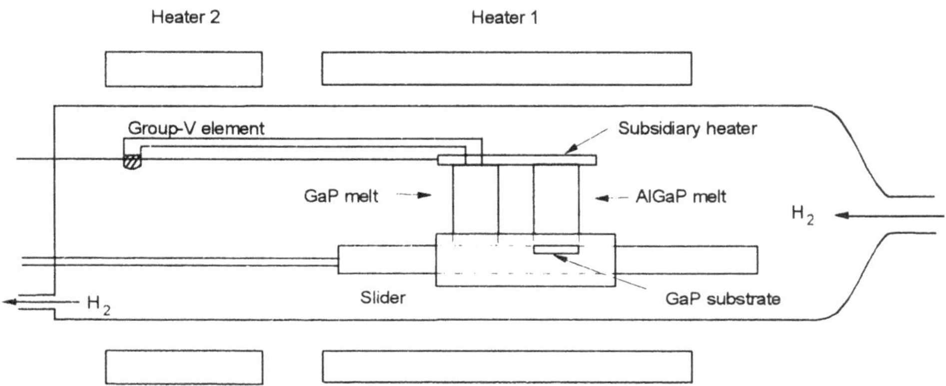

Figure 6.9 Schematic diagram of the growth apparatus for successive epitaxy of $Al_xGa_{1-x}P$.

6.2.2 Lattice Mismatching

For heteroepitaxy, lattice mismatching plays an essential role in crystal imperfectness. Also, we have shown the importance of the optical anisotropy induced by strain on the Raman laser characteristics. The difference in lattice constants of AlP and GaP is $(a_{\mathrm{AlP}} - a_{\mathrm{GaP}})/a_{\mathrm{GaP}} = 3.5 \times 10^{-3}$, as was discussed in Section 5.1.2 [11]. This value is comparable with that of the AlAs-GaAs system (i.e., $(a_{\mathrm{AlAs}} - a_{\mathrm{GaAs}})/a_{\mathrm{GaAs}} = 1.6 \times 10^{-3}$), but it is about one order of magnitude smaller than the values of all other combinations between binary III–V compounds. This is an important feature of the AlP-GaP system. Even so, it has been found that a small lattice misfit causes considerable strain effects.

Figure 6.10 shows surface morphologies of $\mathrm{Al_{0.1}Ga_{0.9}P}$ single epitaxial layers, which reveal cross-hatch patterns caused by generation of misfit dislocations [12]. In Figure 6.10(a), lines are observed both in the $\langle 110\rangle$ and $\langle 1\bar{1}0\rangle$ directions. However, line patterns in one direction are more frequently observed in Figure 6.10(b).

It was made clear both theoretically and experimentally that the differential strain becomes larger at the periphery region of a thin crystal, which generates misfit dislocations when it reaches a value close to half of the lattice spacing, as long as activation energy for generating misfit dislocation can be ignored [13].

Let the mean line spacing be l and the lattice parameter mismatch be Δa, then $\Delta l = l \times \Delta a/a$ can be a measure of a maximam differential strain because Δl corresponds to a maximum value of the difference in the lateral positions of lattice points between two layers just above and below the interface.

The mean line spacing in Figure 6.10(a) is approximately 10 μm. Then, Δl is calculated to be approximately 35Å, higher than the values found in the GaAs—AlGaAsP system (P is added to compensate strain) ranging 3Å to13Å [12]. This may be due to a higher activation energy for generating misfit dislocations in $\mathrm{Al}_x\mathrm{Ga}_{1-x}\mathrm{P}$—GaP heterojunctions.

Asymmetric line patterns may have been caused by larger differential strain at the periphery of a wafer in one direction rather than another, although the reason is not clear.

Neither the lattice constant nor the linear thermal expansion coefficient of AlP at the growth temperature is known. However, as far as comparing linear thermal expansion coefficients α at 300K are concerned, $\alpha(\mathrm{GaP}) - \alpha(\mathrm{AlP}) \approx 0.12 \times 10^{-6}$ K^{-1}, which is about an order of magnitude smaller than $\alpha(\mathrm{GaAs}) - \alpha(\mathrm{AlAs}) \approx 1.1 \times 10^{-6}\ \mathrm{K}^{-1}$. Therefore, accidental coincidence of the lattice parameters would not occur in the GaP—AlP system at the growth temperature, although this does happen in the GaAs—AlAs system. However, this fact does not mean that misfit dislocations are more easily generated in the GaP—AlP system because they will be more easily generated by the temperature change in the difference in the lattice constants, $d\Delta a/\mathrm{d}T$, during the temperature lowering process than they are generated during growth, because the misfitting is small. In fact, we have found larger values of the critical

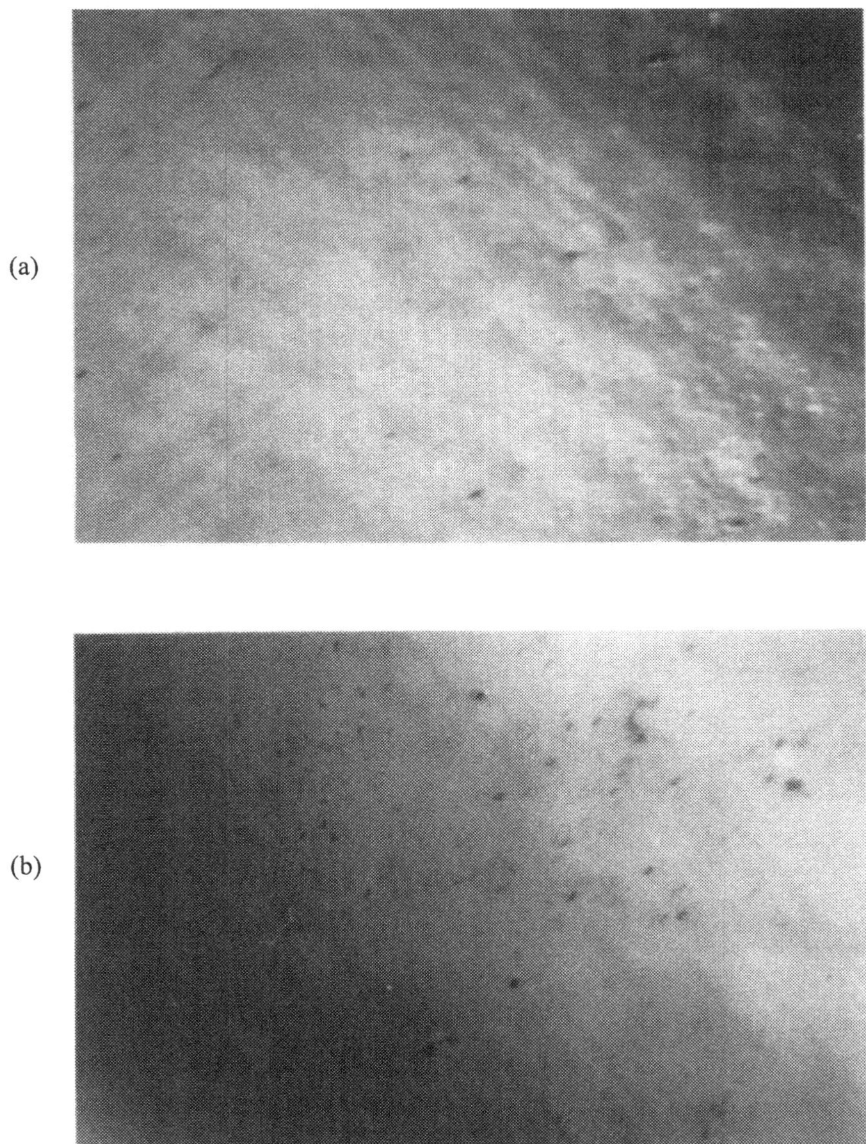

Figure 6.10 Cross-hatch patterns of as-grown $Al_{0.1}Ga_{0.9}P$ epitaxial layers on GaP substrates, thickness (a) 6 μm and (b) 4 μm. Marker represents 100 μm [12].

differential strain for misfit dislocation generation in $Al_xGa_{1-x}P$—GaP interfaces than for GaAs—$Al_xGa_{1-x}As_{1-y}P_y$ interfaces.

The magnitude of residual strain in the substrate and the epitaxial layers can be estimated by measuring optical anisotropy, as was discussed in Section 5.1.2 [11, 12]. A laser beam (620 nm) with a polarization direction 45° from the horizontal is introduced at a height d from the bottom of a wafer, and the depolarization ratio $r = I_{\perp}/(I_{\|} + I_{\perp})$ is measured as a function of d. The measurement for a GaP wave-guiding layer is made by focusing the light beam using an objective lens with f = 10 mm and a numerical aperture N_A = 0.5, while the measurement in the substrate is made by slightly focusing the laser beam with a f = 20 ~ 30 cm lens.

As shown in Figure 6.11(a), the strain induced optical anisotropy gradually increases with increasing height d, with a phase change δ as large as π being reached for wafers having $Al_xGa_{1-x}P$ layers with thicknesses approximately 10 μm.

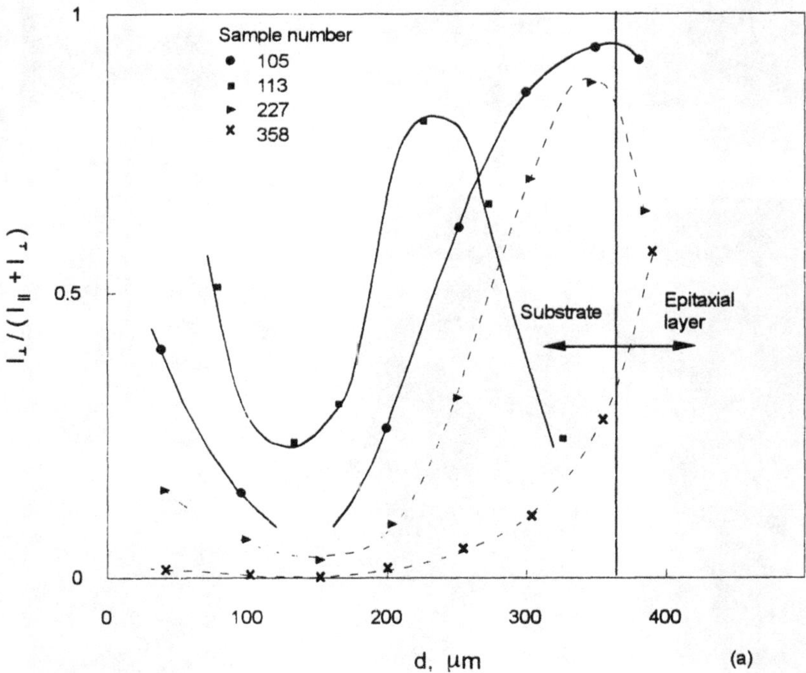

Figure 6.11 (a) Depolarization factor $I_{\perp}/(I_{\|} + I_{\perp})$ as a function of the height d from the bottom of a substrate. Samples 105 and 113 have three layers without stripe structure, while samples 227 and 358 have stripes with 200 μm and 30 μm width, respectively. Aluminum composition x_i and epitaxial layer thickness d_i of sample 105 are $x_1 = x_3 = 0.08$, $d_1 = 3$, $d_2 = 35$, $d_3 = 15$ μm. For sample 113 $x_1 = 0.32$, $x_3 = 0.18$, $d_1 = 2$, $d_2 = 18$, $d_3 = 8$ μm. (b) Lateral strain values [12].

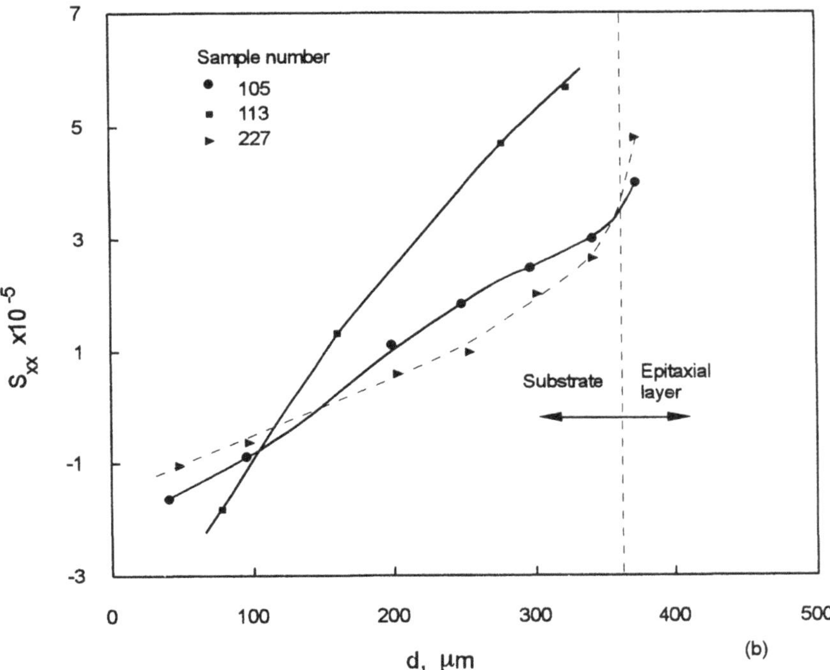

Figure 6.11 (Continued)

We have discussed the relation between the phase change and the strain in Section 5.1.2. When the wafer form is symmetric in the lateral direction, the strain in the lateral direction S_{xx} is related to the phase change as follows:

$$I_{\perp}/(I_{\parallel} + I_{\perp}) = \sin^2 \frac{1}{2}\, \delta.$$

$$\delta = \frac{2\pi}{\lambda} L\Delta\,(n_h - n_v), \quad \Delta(n_h - n_v) \doteqdot 0.13 \times \frac{n^3}{2} S_{xx}. \tag{6.5}$$

Figure 6.11(b) shows strain values calculated in this way. The residual strains on the top surfaces and epitaxial GaP layers are as large as $(3 \sim 5) \times 10^{-5}$. In the case of the buried heterostructure, the lateral strains S_{xx} and S_{yy} are generally different, but the strain tentatively calculated with (6.5) is shown together for comparison (sample 227).

6.3 FREE CARRIERS AND DEEP LEVELS IN THE GROWN LAYERS

It has been shown that the dominant loss mechanisms in GaP Raman lasers are free-carrier absorption and deep-level absorption. As mentioned in Section 3.2, the absorption coefficient at $\lambda = 1\ \mu m$ is in proportion to the free-carrier concentration in n-type GaP, given by $\alpha \simeq 0.05 \times (N/10^{16})\ cm^{-1}$, where $N\ (cm^{-3})$ is the free-carrier concentration [14]. For example, when $N = 2 \times 10^{16}\ cm^{-3}$ and the length $\ell = 5$ mm, the loss for a single path becomes $L = 5\%$, which already exceeds the resonator mirror loss. Therefore, it is desirable that N is much less than $2 \times 10^{16}\ cm^{-3}$. Although the free-carrier concentrations of stoichiometry-controlled GaP epitaxial layers are considerably smaller than those of GaP grown without stoichiometry control, they are still in the range $2 \times 10^{15} \sim 2 \times 10^{16}\ cm^{-3}$.

It is supposed that residual impurities like silicon and oxygen, as well as carbon, are incorporated in the epitaxial layers because the furnace is composed of a quartz tube and a carbon boat. Thus, the free-carrier concentration of an unintentionally-doped epitaxial layer depends on the growth temperature. As shown in Figure 6.12, the conductivity type of GaP is usually n-type, with N in the range $1 \times 10^{16} \sim 2 \times 10^{15}\ cm^{-3}$ when the growth temperature $T_g < 870°C$; while, for $T_g > 870°C$, the conductivity tends to be p-type. On the other hand, the conductivity of $Al_xGa_{1-x}P$ ($x \geq 0.15$) is p-type with hole concentrations $P \simeq 2 \times 10^{16}\ cm^{-3} \sim 2 \times 10^{17}\ cm^{-3}$, when $T_g < 870°C$; while, for $T_g > 870°C$, the conductivity of $Al_xGa_{1-x}P$ tends to be n-type.

Figure 6.13 shows that the hole concentration of $Al_xGa_{1-x}P$ increases with increasing x [15].

On the other hand, deep levels contribute to the loss in the Raman laser by direct absorption and also through the increase of free carriers excited by a high-intensity pump light.

Deep-level concentrations in semiconductors can be obtained by photocapacitance measurement [5,16]. A metal semiconductor contact or a pn-junction of a semiconductor is illuminated by monochromatic light. Then, the capacitance of the depletion layer changes as charges in deep-levels change by excitation of electrons from the deep-levels to the conduction band or holes to the valence band. The capacitance change $\Delta\ C(t)$ at time t saturates to an asymptotic value $\Delta\ C(\infty)$ at low temperature. When $\Delta C(\infty)$ is much smaller than the capacitance of the depletion layer, C, the deep-level concentration is approximately given by:

$$\Delta N_t = \frac{4(V_{bi} - V)}{\varepsilon q A^2} C \cdot \Delta C \tag{6.6}$$

where V_{bi} and V are the built-in and applied voltage, and A is the area of the junction.

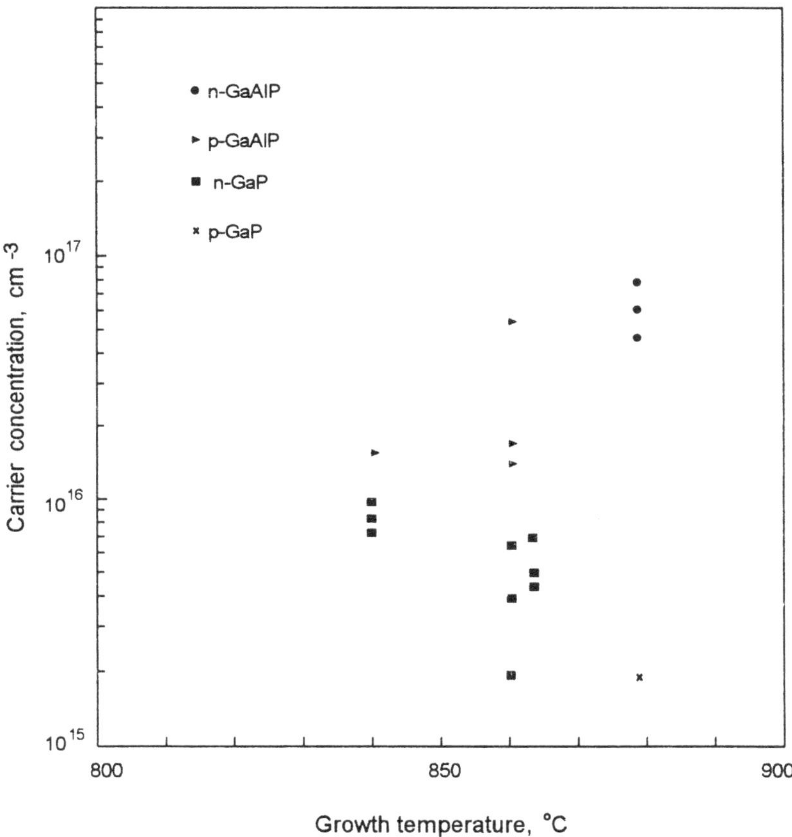

Figure 6.12 Carrier concentrations and conductivity-types of epitaxial GaP and $Al_xGa_{1-x}P$ layers as a function of growth temperature.

The deep-level concentration in GaP was measured for pn-junction light emitting diodes as shown in Figures 6.5 and 6.14 [3]. In this case, the *p* and *n* regions of the pn-junction are doped with shallow impurities Zn and Te, respectively. It is seen that the deep-level concentration becomes a minimum at an optimum phosphorus vapor pressure $P \approx 75$ Torr. We have also measured photocapacitance of the metal-semiconductor contacts of undoped *n*-type GaP epitaxial layers grown with the optimum vapor pressure.

The deep-level concentration is on the order of $10^{14} \sim 10^{15}$ cm^{-3}, as shown in Figure 6.15 [17]. Deep-level concentration is often lower at a facet region than at

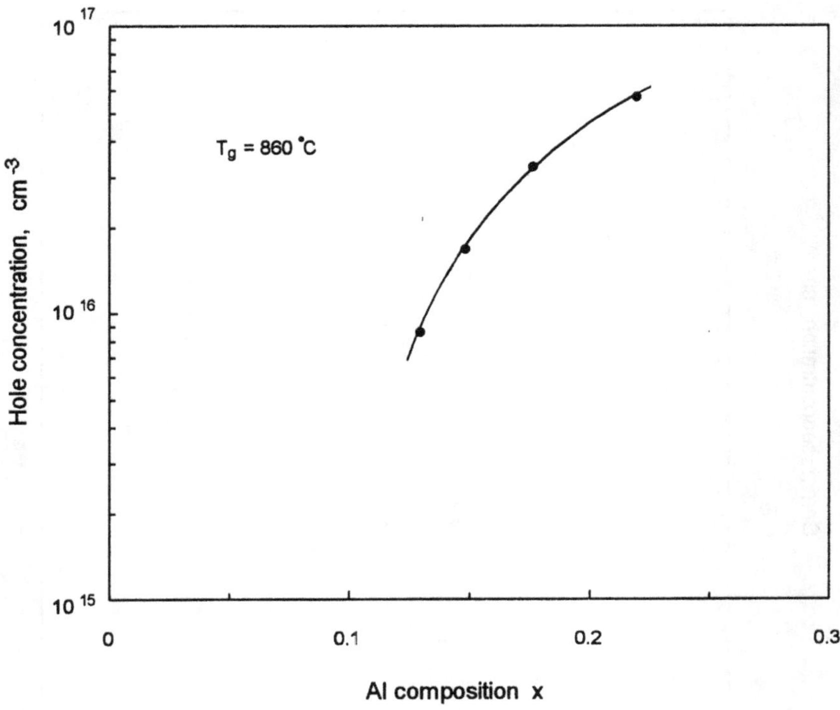

Figure 6.13 Hole concentration of an $Al_xGa_{1-x}P$ cladding layer as a function of aluminum composition x; growth temperature is 860°C [15].

a nonfacet region when compared to those with the same shallow impurity concentration. However, it should be noted that the shallow impurity concentration tends to be lower at a nonfacet region.

The photon energy dependence of photocapacitance can be a measure of the pump wavelength dependence of the loss arising from deep levels in a Raman laser, although they do not directly mean the absorption coefficient. For example, we can say that the number of deep levels responding to a pump light with $\lambda = 1.0\ \mu m$ (1.24 eV) is a few times smaller than the number responding to a pump light with $\lambda = 850$ nm (1.46 eV).

As another point, it should be noted that deep-level concentration tends to increase in proportion to the shallow donor concentration, as shown in Figure 6.16 [9]. This fact means that defect formation is induced by the presence of shallow impurities, even in stoichiometry-controlled crystals. Therefore, the concentration of residual impurities should be kept as low as possible.

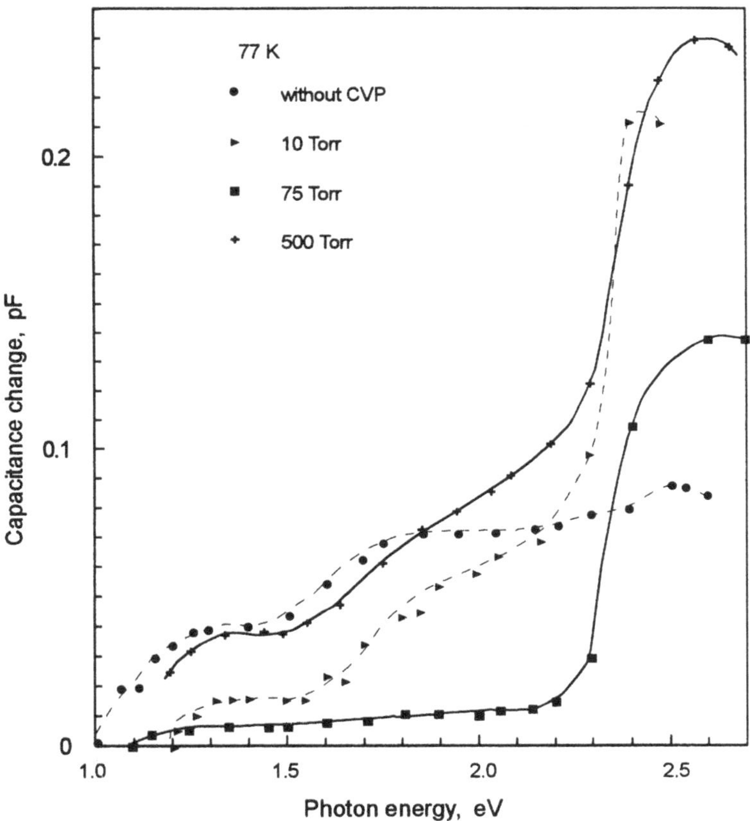

Figure 6.14 Typical photocapacitance spectra of epitaxial GaP layers grown under different phosphorus pressures [3].

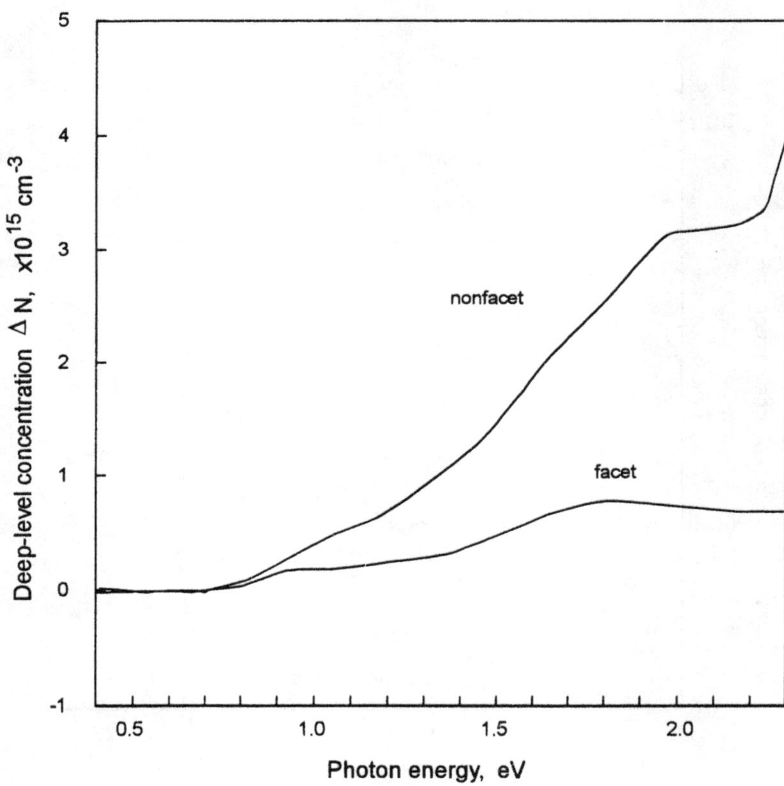

Figure 6.15 Typical photocapacitance spectra of GaP epitaxial layers in nonfacet and facet regions [17].

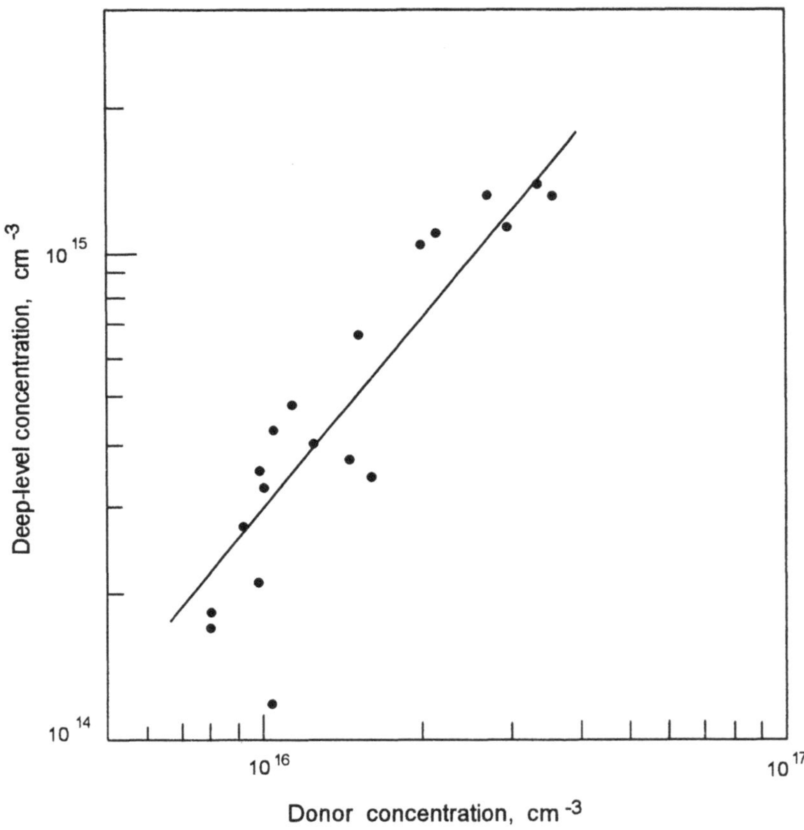

Figure 6.16 Relation between deep-level concentration and shallow donor concentration in GaP epitaxial layers [9].

6.4 DRY ETCHING TECHNIQUE FOR THE WAVEGUIDE FABRICATION

Although the following discussion is concentrated on the fabrication of waveguides for the Raman lasers, the same technique can be applied to the fabrication of waveguides for light modulators as well as light mixers.

The buried waveguides of the Raman lasers are fabricated using the following procedure: (1) the first $Al_xGa_{1-x}P$ cladding layer and the GaP active layer are grown; (2) narrow stripes with width of approximately 5 to 10 μm are formed using the RIE technique; and (3) finally, the second $Al_xGa_{1-x}P$ cladding layer and the GaP overlayer are grown. For the formation of the stripes, the etched depth should be as deep as 4 to 10 μm, while the undercut etching should be kept less than 1 μm, not to degrade the straightness of a stripe.

The RIE technique can be employed for deep vertical etching of GaP—$Al_xGa_{1-x}P$ layers with small undercut. Various etching gases have been reported for GaAs and other III–V compound semiconductors, as is given in Table 6.1 [18].

We employed PCl_3 gas for etching GaP—$Al_xGa_{1-x}P$ [19], rather than Cl_2, which is most often used for etching of GaAs. PCl_3 is easier to handle than Cl_2 because it is less reactive, and it is in a liquid state at around room temperature. As shown in Table 6.2, the boiling point of PCl_3 is 76°C. Table 6.2 also gives the boiling points of various halogen compounds relating to etching reactions [20]. In the etching reaction of GaP and $Al_xGa_{1-x}P$ by PCl_3 gas, the reaction products are thought to be $GaCl_3$, $AlCl_3$, and PCl_3, as well as P_2 or P_4. It should be noted that both the boiling points of $GaCl_3$ and $AlCl_3$ are relatively low so that reaction products are easily dissociated from the crystal surface.

Figure 6.17 shows the schematic diagram of an RIE apparatus. A glass chamber containing PCl_3 liquid is maintained at 0°C, and the gas pressure of PCl_3 introduced into a reaction chamber is controlled between 0.01 to 0.05 Torr by using a needle valve. The plane parallel electrodes in the reaction chamber are coupled to an RF generator operating at frequency 3.2 MHz. The power level of the RF generator is usually 400W.

The sample crystal is placed on a quartz pedestal that covers the cathode electrode.

If we use a single layer of photoresist as a mask in RIE etching, a sizable deterioration of the stripe patterns occurs. Instead, we use a double layer made of Si_3N_4, deposited by CVD, and positive photoresist OFPR 800 baked at 145°C. As a result, the lateral undercut is reduced to less than 0.5 μm.

The etching rate as well as the etched profile depend on the PCl_3 gas pressure. Figure 6.18 shows the etched profiles at three different gas pressures. At the highest gas pressure $P = 0.031$ Torr, the etched wall is far from vertical. When the gas pressure is lowered to 0.019 Torr, the etched profile is improved, although the etching rate is reduced to approximately 1 μm/min. At a further lower gas pressure, discharging becomes unstable. Although we have used a conventional RIE apparatus,

Table 6.1
Etching Gases and Etching Rates for III–V Compound Semiconductors*

RIE : Reactive ion etching
RIBE : Reactive ion beam etching
IBAE : Ion beam assisted etching
RE : Radical etching
PE : Plasma etching

Crystal	*Gas*	*Etching Method*	*Etching Rate*
GaAs	$CCl_2F_2(O_2, Ar, He)$	RIE	0.2–1.0 μm/min
	$CCl_4(O_2, H_2, Cl_2)$	RIE	4–10
		RIBE	0.3–0.6
	$CCl_3(O_2)$	RIE	4–13
	Cl_2	RIE	
		RIBE	6
		RE	0.1–0.6
	Cl_2/Ar^+	IBAE	1–10
	$Cl_2(Ar)$	RIE	0.1–0.8
	$Cl_2/Cl^+/Cl_2^+$	RIBE	1–5
	HCl	RIE	0.05–0.1
	$BCl_3(Cl_2)$	RIE	0.06
	Br_2	RE	0.2–1.2
	H_2	RE	25–70
	$SiCl_4$	RE	0.02–0.06
AlGaAs	$CCl_2F_2(He)$	RIE	0.1–0.6
	$Cl_2(Ar)$	RIE	0.03–0.2
	Cl_2	RIBE	0.3–1.0
	$BCl_3(Cl_2)$	RIE	0.1–0.6
InP	$CCl_2F_2(Ar, O_2)$	RIE	0.1–0.8
	$Cl_2(O_2)$	RIE	0.1–0.25
	$CCl_4(O_2, Cl_2)$	RIE	1.0–1.5
	$Cl_2/Cl^+/Cl_2^+$	RIBE	1.4–2.2
	$CCl_3F(O_2)$	RIE	0.05–0.08
InGaAsP	$Cl_2(O_2, Ar)$	RIE	4.0–18
	$CCl_4(O_2)$	RIE	
GaAs-Oxide	CCl_4	RIE	
	CCl_2F_2	RIE	
	PCl_3	RIE	
	HCl	RIE	
	H_2	RIE	
			0.03–0.12

*After [18].

Table 6.2
Boiling Points of Halide Materials*

Material	*Boiling Point (°C)*	*Material*	*Boiling Point (°C)*
AlF_3	1,291	$InBr_2$	632
$AlCl_3$	183	PF_5	−75
$AlBr_3$	263	PF_3	−101
GaF_3	1,000	PCl_5	162
$GaCl_2$	535	PCl_3	76
$GaCl_3$	201	PBr_5	106
$GaBr_3$	279	PBr_3	173
InF_3	1,000	AsF_5	−53
InCl	608	AsF_3	−63
$InCl_2$	560	$AsCl_3$	130
$InCl_3$	600	$AsBr_3$	221

*After [20].

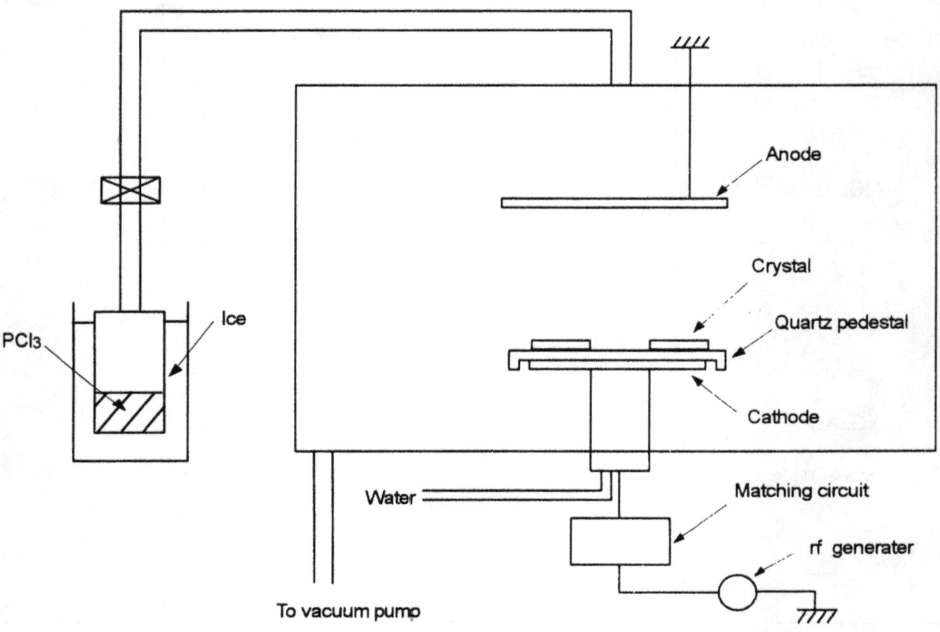

Figure 6.17 Schematic diagram of a reactive ion etching (RIE) apparatus.

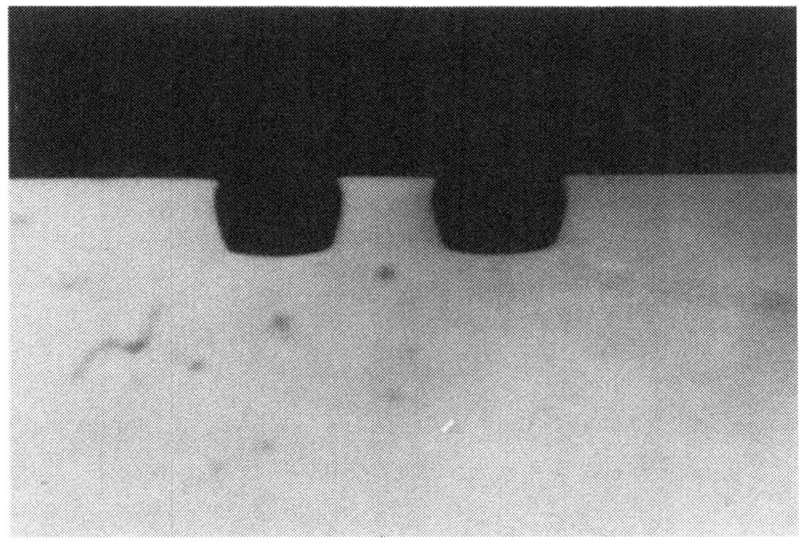

(a)

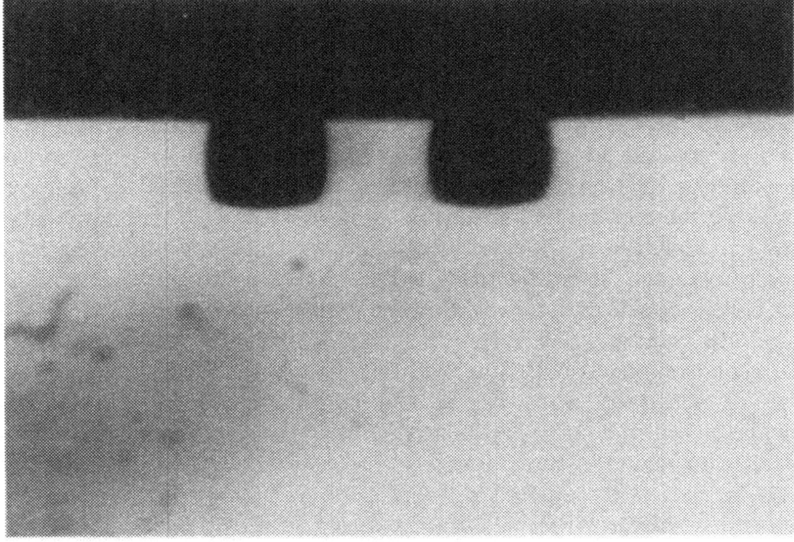

(b)

Figure 6.18 Profiles of GaP stripes formed by the RIE process using PCl_3 Gas: (a) $P(PCl_3) = 0.031$ torr, (b) 0.025 torr, and (c) 0.019 torr.

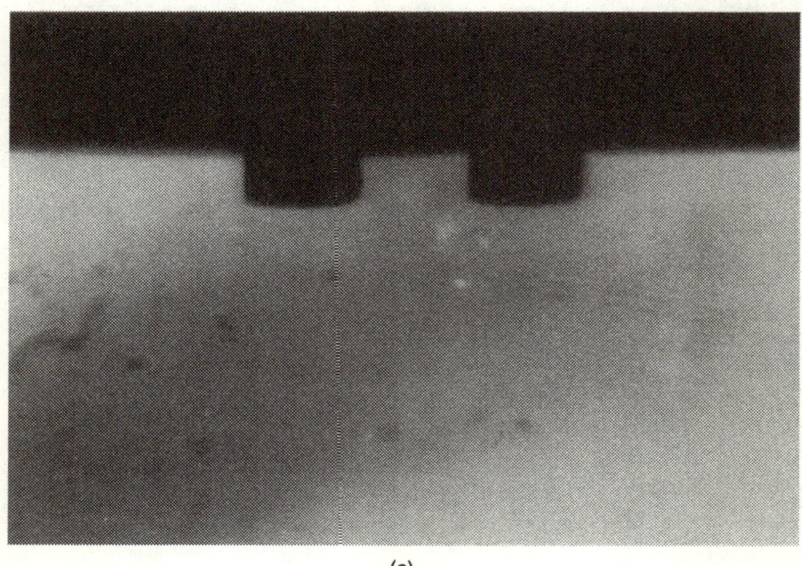

(c)

Figure 6.18 (Continued)

the electron cyclotron resonance plasma method will give a better result because the gas pressure and the kinetic energies of particles can be reduced.

Asakawa showed an excellent etching profile of GaAs—$Al_xGa_{1-x}As$ by employing an electron cyclotron resonance plasma using Cl_2 gas [21]. He used a trilevel resist mask, the top and the bottom layers being AZ-1350J and the middle layer an evaporated titanium film. As a result, deep and remarkably straight vertical etched patterns were obtained, having a stripe width/groove width = 0.5 μm/2 μm and 1.0 μm/1.0 μm, and with 7-μm depth. The optimum gas pressure was as low as between 5×10^{-4} and 1×10^{-3} Torr.

The second problem of RIE, as well as electron cyclotron resonance, is a rippled surface of etched side walls. Figure 6.19 shows a scanning electron micrograph of a sidewall surface of a stripe. Although its origin is not understood at present, it may be related to lateral etching. It is also not known how much the rippled side wall affects the loss of the guided light wave.

The third problem of RIE is surface damage produced by the irradiating particles. In the case of GaAs, it was shown by Rutherford back scattering that the thickness of a highly damaged amorphous layer at the surface is only approximately 2 to 7 nm. However, it was found from the decrease of photoluminescence intensity that an imperfect crystalline layer where defects are diffused in extends over 200 nm [22].

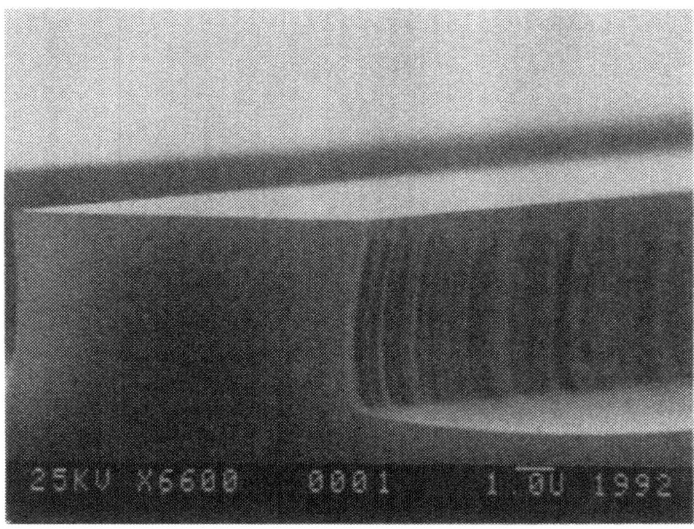

Figure 6.19 SEM photograph of a side wall of a GaP stripe. $P(PCl_3)$ = 0.019 torr. Etching time, 4 min.

Excessive numbers of deep levels present at the waveguide interface cause a serious effect on the lasing performance. Therefore, at the very least, the heavily damaged surface layer must be removed after RIE processing. Although light wet etching by a conventional solution like HF : HNO_3 : H_2O = 1 : 1 : x ($x = 0 \sim 4$) or HCl : HNO_3 : H_2O = 1 : 3 : x ($x = 0 \sim 4$) was thought to be effective, it was found that these solutions caused a rugged pattern at the corners of stripes, as shown in Figure 6.20.

On the contrary, a solution of H_2SO_4 : H_2F_2 : H_2O = 4 : 1 : 1 was found to cause no such irregular etching, provided the etched depth is 60 nm or less. We employed this etchant for the fabrication of the Raman laser working at a 1.0-μm region. However, the Raman lasers working at the wavelength region 840 to 900 nm revealed serious effects of deep levels at the waveguide interface, as discussed in Section 5.4.

We have found that a solution of methyl-alcohol : Br_2 = 100 ~ 200 : 1 causes smooth etching up to a lateral etching depth of 1 μm. Figure 6.21 shows a SEM photograph of an after-etched surface of a stripe fabricated by RIE.

Figure 6.22 shows that the etching rate depends on Br_2 concentration and temperature of the solution. It should be noted that the etching rate at the stripe region is about 5 times larger than the etching rate for a plane surface of GaP crystals. It should also be noted that the etching rate for $Al_xGa_{1-x}P$ ($x \simeq 0.1 \sim 0.2$) is larger than that for GaP.

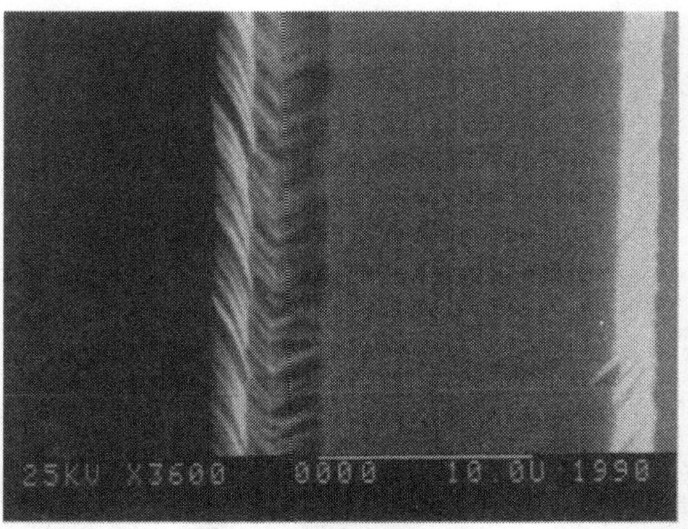

Figure 6.20 Top view (SEM) of a GaP stripe after light etching process by ($HF : HNO_3 = 1:1$) solution. $P(PCl_3) = 0.031$ torr.

Figure 6.21 SEM profile of a GaP stripe after light etching process by ($Br_2 : CH_3OH = 1:100$) solution. $P(PCl_3) = 0.019$ torr.

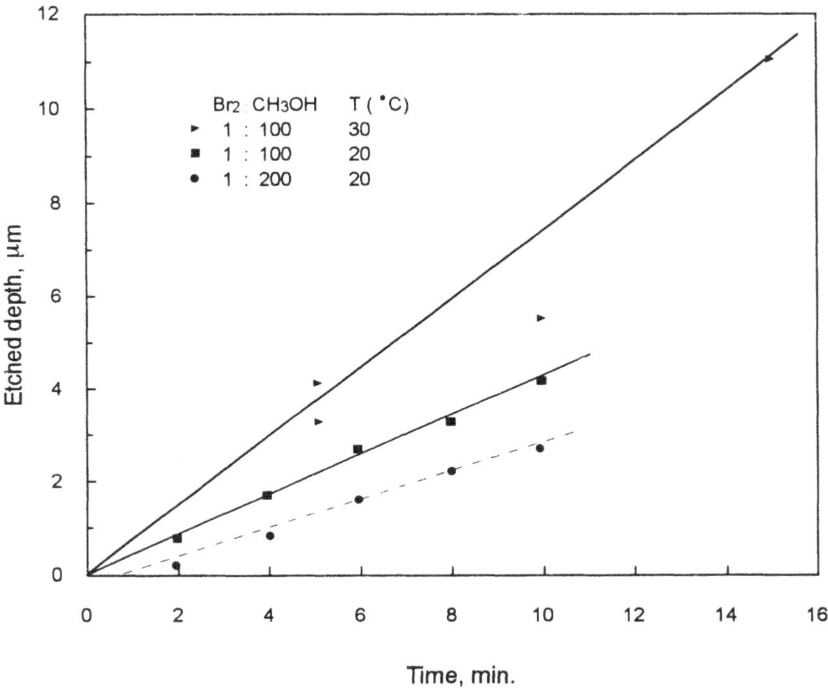

Figure 6.22 Etching rate of GaP by Br_2-methanol solution.

REFERENCES

[1] J. Nishizawa and Y. Okuno, "Liquid Phase Epitaxy by a Temperature Difference Method Under Controlled Vapor Pressure," *IEEE Trans. Electron Devices*, ED22, 1975, pp. 716–721.

[2] J. Nishizawa, Y. Okuno and H. Tadano, "Nearly Perfect Crystal Growth in III–V Compounds by the Temperature Difference Method Under Controlled Vapor Pressure," *J. Crystal Growth*, Vol. 31, 1975 , pp. 215–222.

[3] J. Nishizawa and Y. Okuno, "Stoichiometric Crystallization Method of III–V Compounds for LED's and Injection Lasers," *Semiconductor Optoelectronics*, ed. M. A. Herman, PWN-Polish Scientific Publishers, Warsaw, 1980, pp. 101–130.

[4] J. Nishizawa, Y. Okuno and K. Suto, "Nearly Perfect Crystal Growth in III–V and II–IV Compounds," JARECT 19, *Semiconductor Technologies*, ed. J. Nishizawa, OHMSHA Ltd and North-Holland, Tokyo, 1986, pp. 1–80.

[5] J. Nishizawa, "Stoichiometry Control of Compound Semiconductor Crystals," *Non-Stoichiometry in Semiconductors*, ed. K.J. Bachmann, H.L. Hwang and C. Schwab, Elsevier Science Publishers, 1992, pp. 95–106.

[6] J. Nishizawa and H. Sakuraba, "Surface Reaction Mechanism in Si and GaAs Crystal Growth," *Surface Science Reports*, Vol. 15,1992, pp. 137–204.

[7] R.E. Honig and D.A. Kramer, "Vapor Pressure Data for the Solid and Liquid Elements," *RCA Rev.*, 1969, pp. 285–305.

[8] J. Nishizawa, Y. Okuno, M. Koike, and K. Nishibori, "Effects of Vapor Pressure on GaP LED's," *Jpn. J. Appl. Phys.*, Vol. 17, Supp. 17–1, 1977, pp. 87–92.

[9] J. Nishizawa, Y.J. Shi, K. Suto and M. Koike, "Photocapacitance Study of Deep Levels Due to Nonstoichiometry in Nitrogen-Free GaP Light-Emitting Diodes," *J. Appl. Phys.*, Vol. 53, 1982, pp. 3878–3883.

[10] A. Tanaka, T. Sugiura and T. Sukegawa, "Low Temperature Phase Diagram of Ga—Al—P Ternary System," *J. Crystal Growth,* Vol. 60, 1982, pp. 120–122.

[11] K. Suto, T. Kimura and J. Nishizawa, "Heterostructure Semiconductor Raman Laser," *IEE PROC.*, Vol. 134, Pt. J, 1987, pp. 215–220.

[12] K. Suto, T. Kimura, S. Ogasawara and J. Nishizawa, "Heteroepitaxy of GaP—$Al_xGa_{1-x}P$ System by the Temperature Difference Method Under Controlled Vapor Pressure (TDM-CVP)," *J. Crystal Growth*, Vol. 99, 1990, pp. 297–301.

[13] J. Nishizawa, Y. Okuno, M. Fukase, and H. Tadano,"Lattice Strain and Misfit Dislocation in GaAs—GaAlAsP Heterojunctions," *J. Crystal Growth*, Vol. 52, 1981, pp. 929–935.

[14] K. Suto and J. Nishizawa, "Semiconductor Raman Laser," *IEE PROC.*, Vol. 132, Pt. 3, 1985, pp. 81–84.

[15] K. Suto, T. Kimura and J. Nishizawa, "Semiconductor Raman Laser Pumped With a Fundamental Mode," *IEEE PROC. J.*, Vol. 139,1992, pp. 407–412.

[16] J. Nishizawa, M. Koike, K. Dezaki and J. Shibata, "Automatic Measurement System for Photocapacitometry Analysis," *Rev. Sci. Instrum.*, Vol. 57, 1986, pp. 453–462.

[17] K. Suto, D. Iwamoto, J. Nishizawa and N. Chubachi, "Free Exciton Recombination in Stoichiometry-Controlled GaP," *J. Electrochem. Soc.*, Vol. 140, 1993, pp. 2682–2687.

[18] K. Asakawa, *Dry Etching of GaAs* (in Japanese), Handoutai-Kenkyu (Semiconductor Research) Vol. 25, 1986, pp. 255–298.

[19] K. Suto, T. Kimura, and J. Nishizawa, "Lateral Optical Confinement of the Heterostructure Semiconductor Raman Laser," *Appl. Phys. Lett.*, Vol. 51, 1987, pp. 1457–1458.

[20] R.C. Weast and M.J. Astle, *CRC Handbook of Chemistry and Physics*, Boca Raton, Florida: CRC Press, 1980, B-73.

[21] K. Asakawa and S. Sugata, "GaAs and AlGaAs Anisotropic Fine Pattern Etching Using a New Reactive Ion Beam Etching System," *J. Vac. Sci. Technol.*, Vol. B3, 1985, pp. 402–405.

[22] M. Kawabe, N. Kanzaki, K. Masuda and S. Namba, "Effects of Ion Etching of the Properties of GaAs," *Appl. Opt.*, Vol. 17, 1978, pp. 2556–2561.

Chapter 7
Amplification and Demodulation in Wideband Optical Communication

In long-distance fiber communication systems, light amplifiers are of great importance. Figure 7.1 shows that there are possibly four kinds of light amplifier applications. One is the postamplifier, which amplifies the light output power from the laser diode signal source. The second is the inline repeater amplifier, which amplifies the light signal at an intermediate station of a fiber communication system. The third is the receiver preamplifier, which is used for a demodulation system. The fourth is called a remotely pumped amplifier. This is a piece of erbium-doped fiber located several tens of kilometers from the end of a system, pumped by a laser located in the terminal station on shore. Using a remotely pumped amplifier, the system reach can be extended by approximately 40 to 70 km. For the first two, high power is the most important. As for the gain, a single high-gain amplifier, although it is available, can lead to real system problems in the absence of a data signal. Therefore, it is desirable to use multiamplifier links where amplifiers are used in compression with each amplifier having low to medium gain (7 dB for trans-Pacific systems and 20 dB for 1000-km type links).

On the other hand, the preamplifier should have extremely low noise characteristics rather than high gain or high-power capability. Tuning of frequency is also required so that the preamplifier has selectivity in frequency domain. First, we discuss the erbium-doped fiber amplifier, which is used as an inline repeater amplifier as well as a postamplifier. The Er^{3+}-doped fiber has excellent amplifier characteristics. Although its amplifying wavelengths are limited to approximately 1.5 μm, they are coincident with the minimum loss wavelength region of the optical fiber. Semiconductor laser amplifiers are also being extensively investigated. In contrast to fiber amplifiers, their available wavelengths are widely spread. However, nonlinear interactions in semiconductor lasers make the amplifier characteristics somewhat

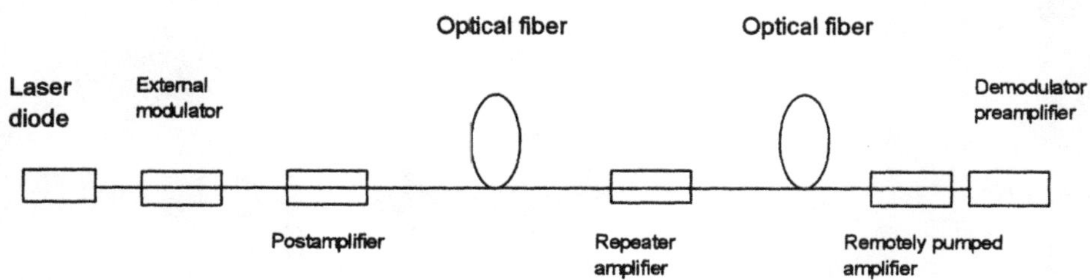

Figure 7.1 Wideband and long-distance optical communication system with optical amplifiers.

complicated. For the preamplifiers, as well as demodulators, we think that the semiconductor Raman laser is the most promising. It is a frequency-selective light amplifier. The semiconductor Raman laser can be used for selective amplification of the desired frequency only. We discuss the heterodyne-type demodulation method based on the Raman laser amplifier in a wideband optical communication system.

7.1 FIBER AMPLIFIERS

7.1.1 Erbium-Doped Fiber Amplifier

Since T. H. Maiman implemented the first laser by using Cr^+-doped ruby crystals, various kinds of transition metal ions and rare-earth metal ions incorporated into insulator crystals, as well as into glasses, have been extensively studied. Figure 7.2 shows lasing wavelengths available from these ions [1]. Although the proposals to use glass fibers for the laser mediums were made as early as the 1960s [2,3], the development of erbium-doped fiber amplifiers was stimulated by the development of the long-distance optical fiber communication in the 1980s [4].

In the Er^{3+}-doped silica glass, light amplification occurs at the 1.5-μm wavelength region, and its emission spectrum has two peaks at 1.552 μm and 1.535 μm. As shown in Figure 7.3, it is in the region of the minimum in the loss-vs-wavelength curve of silica glass fiber [5,6].

On the other hand, Figure 7.4 shows the refractive index dispersion curve of the silica glass fiber [6]. It becomes zero at $\lambda \simeq 1.3$ μm. For high-speed optical communications, zero dispersion is an important property. Therefore, light amplification in the 1.3-μm region is also attracting much attention, and Pr^{3+} and Nd^{3+} doped fibers are being studied. Prototype Pr^{3+} amplifiers are now available with 18-dBm output power and <6-dB noise figure. This is as good as most typical Er^{3+} amplifiers. A drawback is that it needs high pump power of several hundred milliwatts [7,8].

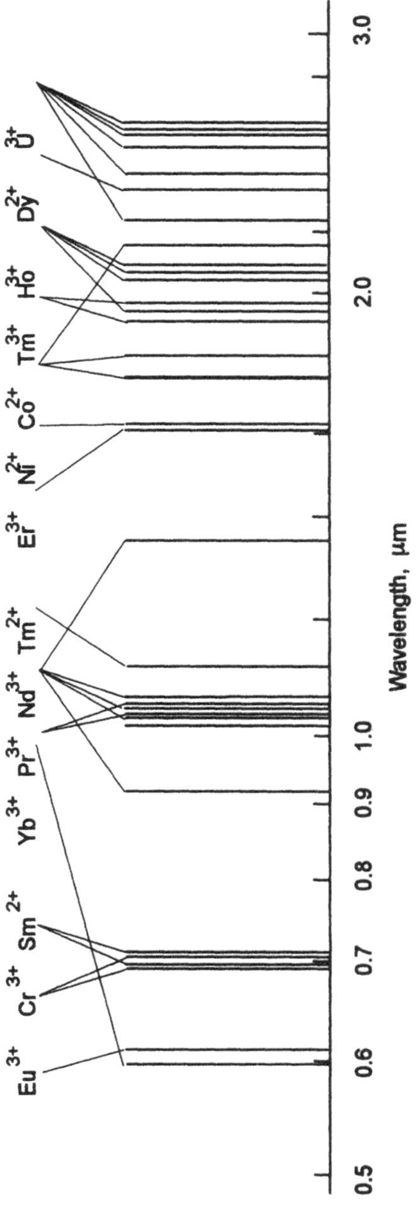

Figure 7.2 Lasing wavelengths of solid-state lasers [1].

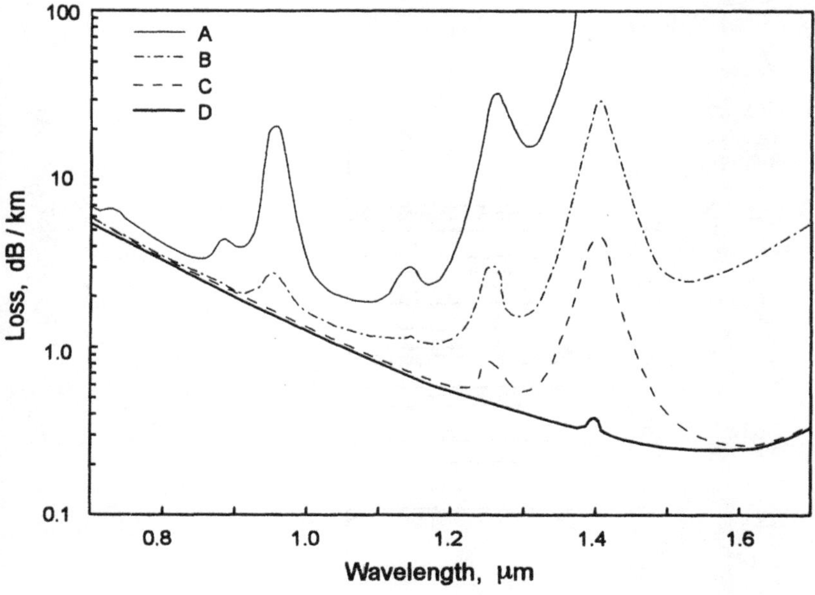

Figure 7.3 Loss spectrum of a completely dehydrated fiber. The lines denoted by A, B, C, and D represent the best data in each year (A, 1977; B, 1978; C, 1979; D, 1980) [5].

Light amplification by Er^{3+} ions can be discussed in terms of a three-level system [3]. Figure 7.5 shows the energy-level diagram of an Er^{3+} ion.

Absorption of pumping light results in various excited electronic levels depending on the pump light wavelength. Then, the excited ions relax through nonradiative processes to a metastable state $^4I_{13/2}$,which we call the upper state. Finally, the ions return to the ground state, with emission of photons. In the absence of the incident photons, the lifetime of the upper state of Er^{3+} is as long as 13 ms, so the population inversion occurs easily. Optical pumping is carried out using a high-power laser diode in practical systems. As shown in Figure 7.5, there are three different pump wavelength regions, 800 nm, 980 nm, and 1.48 μm, for which appropriate laser diodes are available. Figure 7.6 shows the wavelength ranges of various kinds of laser diodes.

The amplification band of an Er^{3+}-doped silica fiber can be determined from the amplified spontaneous emission (ASE) spectrum, as shown in Figure 7.7 [9]. Photons emitted by the spontaneous emission process are amplified along the optical fiber so that the ASE spectrum observed at the output end of the fiber approximately gives the amplification band profile, although both are not identical to each other. As illustrated in Figure 7.7, the 3-dB bandwidth around the 1.535-μm peak is about 250 GHz. The amplifier band is broader than that in a crystalline quartz. This is

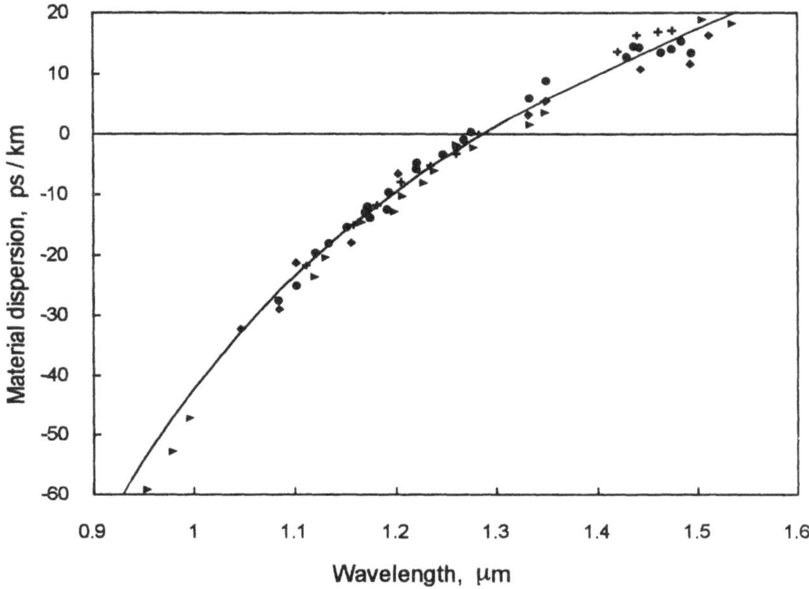

Figure 7.4 Material dispersion of single-mode optical fibers with maximum refractive-index difference $\Delta = 0.265\%$ [6].

because local crystal fields acting on Er^{3+} ions are slightly different from site to site. In spite of such a local field distribution, the amplifier band is found to be homogeneously broadened at room temperature, probably as a result of fast cross-relaxation among the sublevels of $^4I_{13/2}$ and $^4I_{15/2}$ [10,11]. Inhomogeneous nature is observed at lower temperatures (e.g., at 77K). If atoms such as Al are incorporated into silica fiber, the gain bandwidth is even further broadened.

Following the discussion in [12], let us assume first that the input signal light power and the signal output power are so small that the populations in the upper and ground states, N_2 and N_1, can be approximated as remaining highly inverted. Then, the signal power, I_s, is amplified along the fiber length according to:

$$\frac{dI_s}{dz} \simeq (\sigma_s N_2^0 - \sigma_a N_1^0) I_s \tag{7.1}$$

where σ_s is the stimulated emission cross section of the upper state level 2, σ_a is the absorption cross section of the ground state level 1, and N_2^0 and N_1^0 are the populations of both levels under the condition that I_s is small.

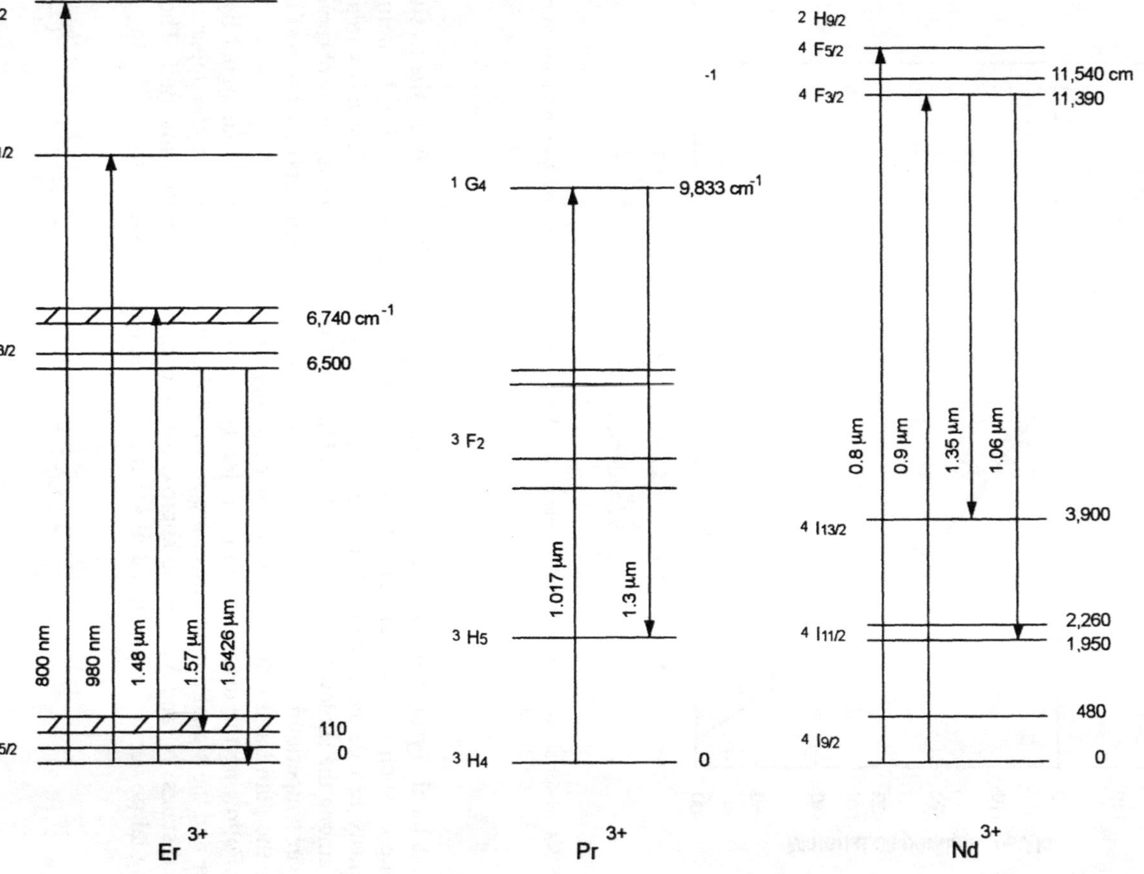

Figure 7.5 Energy level diagrams of Er^{3+}, Pr^{3+}, and Nd^{3+}.

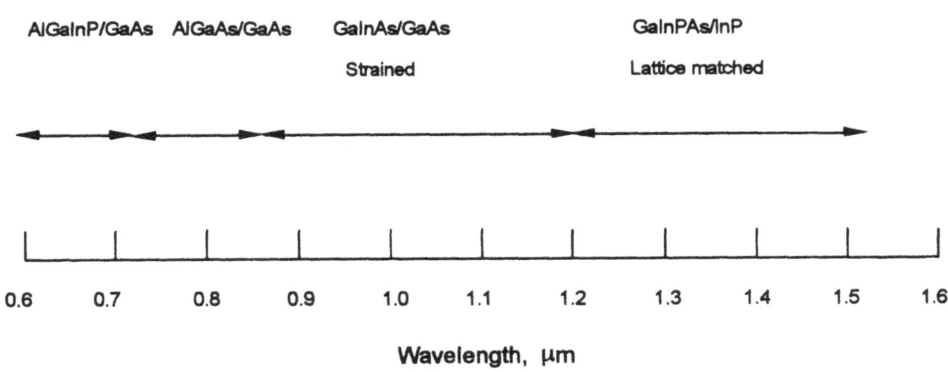

Figure 7.6 Wavelength ranges of various kinds of laser diodes.

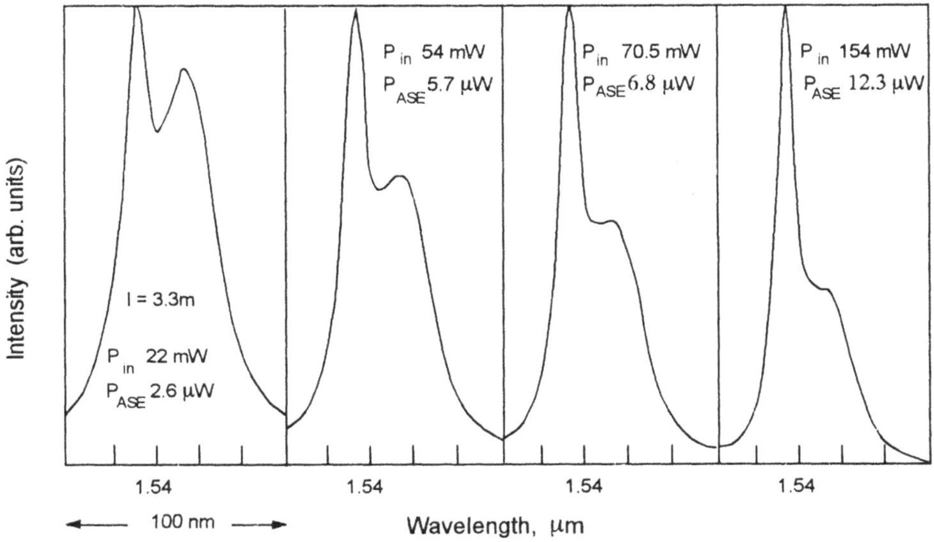

Figure 7.7 ASE spectra obtained around 1.54 μm with a fiber length L = 3.3 m [9].

The gain per unit length is given by:

$$g_0 = \sigma_s N_2^0 - \sigma_a N_1^0 \tag{7.2}$$

where g_0 is a function of a pump power and a pump wavelength.

Then, let us assume that the signal light is modulated with pulses or any other waveform. The input signal electric field E_s can be represented by a sum of Fourier components $E_s(\omega_i)$:

$$E_s = \Sigma E_s(\omega_i)e^{j\omega_i t} \tag{7.3}$$

Amplification equations are usually described in terms of photon density or power density $|E_s|^2$. However, we must describe amplification in terms of the electric field E_s to explicitly express the phase relation. Within a small signal approximation, the amplification coefficient for the electric-field per unit length is given by $g/2$, when gain per unit length is given by g.

Then, each component $E_s(\omega_i)$ is amplified according to:

$$\frac{dE_s}{dz}(\omega_i) = \frac{1}{2}(\sigma_s(\omega_i)N_2^0 - \sigma_a(\omega_i)N_1^0)E_s(\omega_i) \tag{7.4}$$

provided the signal level is far below the saturation of the amplifier.

Therefore, all components are amplified by nearly equal factors, as long as they are within the amplification 3-dB bandwidth so that there is no distortion of the signal waveform.

Then, let us consider when the input signal is increased near to the saturation level. We can no longer assume that N_2 and N_1 remain constant. In fact, N_2 falls while N_1 increases. As was pointed out earlier, the gain band of Er^{3+}-doped fiber is nearly homogeneously broadened, so that spectral hole burning that should drastically change the gain-band profile will not take place. When the gain band is homogeneously broadened, the gain g can be described as:

$$g = g_0(1 + I_s/I_{sat})^{-1} \tag{7.5}$$

where I_{sat} is the signal power level at which the gain is reduced by 3 dB. The actual output signal power cannot increase very much above I_{sat}. The value of I_{sat} increases nearly in proportion to the pump power, I_p, when I_p is well above the threshold value for amplification. Figure 7.8 shows an example of amplifier gain characteristics, which illustrates output power saturation through gain reduction [13].

The next important point about light amplifiers is their noise characteristics. It is obvious that a major source of noise is due to ASE. The ASE power is in proportion to the amplifier bandwidth Δf.

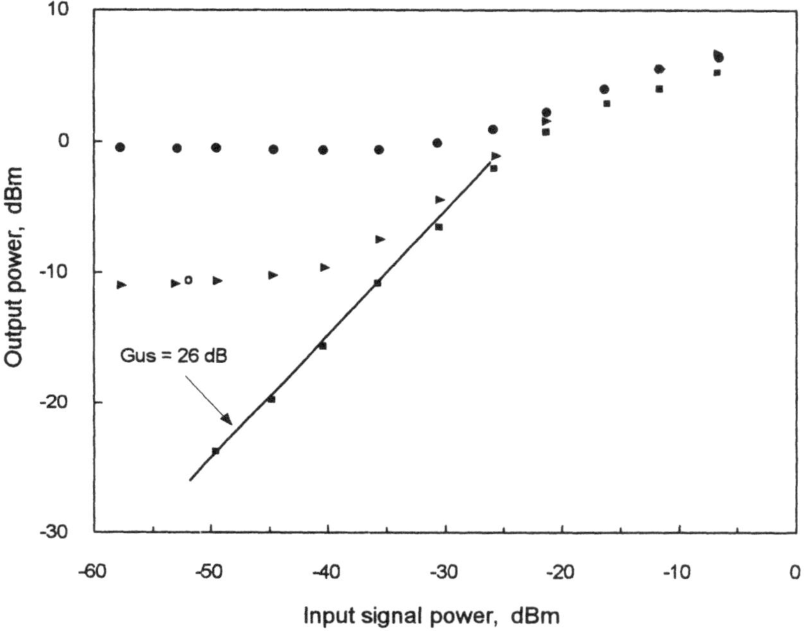

Figure 7.8 Amplification characteristics of a first-stage amplifier of a two-stage amplifier system: ● = without optical filters; ▶ = with optical filter with $\Delta\lambda$ = 1.0 nm; ■ = net output power after the ASE power is subtracted [13].

The noise of the light signal can be considered in terms of photon statistics. We briefly follow the discussion in [12]. The average number of signal photons counted in a unit time $\langle n_{out} \rangle$ and its variance σ^2_{out} can be described in the following form, if we assume that the input signal is perfectly coherent:

$$\langle n_{out} \rangle = G\langle n_{in} \rangle + (G - 1)n_{sp} m\Delta f$$

$$\sigma^2_{out} = \langle n^2_{out} \rangle - \langle n_{out} \rangle^2 = G\langle n_{in} \rangle + (G - 1)n_{sp} m\Delta f + 2G(G - 1)n_{sp} \langle n_{in} \rangle + (G - 1)^2 n^2_{sp} m\Delta f \tag{7.6}$$

where G is the gain, n_{sp} is the spontaneous emission factor which is equal to $\frac{N_2}{N_2 - N_1}$, and m is the number of allowed modes in the fiber. If we use an ideal photodetector, the noise power $N_d(f)$ per unit frequency interval is given by:

$$N_d(f) = 2e^2\sigma_{out}^2 \tag{7.7}$$

In the expression for σ_{out}^2 (7.6), the first and the second terms arise from shot noise due to the signal and ASE, while the third term is due to the beat between the signal and ASE, and the fourth term is due to the beat between ASEs. When the gain is high, as in the case of the Er^{3+}-doped fiber amplifier, the beat noise terms are dominant. It should be noted that the ASE-ASE beat noise, as well as the ASE shot noise, is proportional to the amplifier bandwidth. It should also be noted that if we use the amplifier in the power saturation mode, the spontaneous emission factor n_{sp} increases so that the noise power increases. Actually, the bandwidth of noise reaching the detector is determined by the bandwidth of optical filters placed before the detector. Typical bandwidth of a conventional narrow band filter is $\Delta\lambda \simeq 1$ nm (i.e., $\Delta f \simeq 130$ GHz). Figure 7.8 shows that the ASE noise from an Er^{3+}-doped fiber amplifier is effectively filtered out by a narrow bandpass filter.

The filter bandwidth can be further narrowed by using fiber Fabry-Perot resonators as well as etalons, although tuning of the center frequency becomes a problem. Filters actually benefit amplifier systems by reducing the Gordon-Hauss effect in soliton systems [14]. When the amplifier gain is very high, the backward-traveling ASE causes a gain efficiency decrease and noise figure increase. This effect can be eliminated by inserting an isolator in a two-stage fiber amplifier system [15]. Medium-level amplification of $G \simeq 20$ dB per amplifier is thought to be adequate for a long-distance system over 1,000 km [16].

In conclusion, the Er^{3+}-doped fiber amplifier is the most suitable for an inline repeater, which needs wideband capability, high gain, and high output power rather than low noise characteristics. It is also suitable for the postamplifier, which amplifies the output from the signal source laser diode.

On the contrary, for the preamplifier, which is to be used in the demodulator system, the noise figure and tunability are more important than gain and high-output power, and the semiconductor Raman laser will be the most suitable amplifier, as will be discussed in Section 7.3.

7.1.2 Fiber Raman Amplifier for Soliton Propagation

There are two possible different approaches for future optical communications. One is based on frequency division (i.e., wideband optical communication), while the other is based on time division (i.e., high-speed optical communication).

For the latter, the dispersion of the refractive index of glass finally limits the propagation distance of the optical pulse because the pulse width is inevitably broadened. Soliton propagation in a fiber to overcome the pulse width broadening was proposed by A. Hasegawa and F. Tappert [17]. It is based on the nonlinear property of the refractive index of a fiber in terms of the electric field E of the light. In a medium with inversion symmetry, the refractive index at high field can be represented as:

$$n = n_0(\omega) + i\chi(\omega) + n_2\overline{E^2} \tag{7.8}$$

The last, nonlinear, term represents the optical Kerr effect, while χ represents the dielectric loss. For SiO_2, the relevant values are $n_0 \simeq 1.5$ and $n_2 \simeq 3 \times 10^{-22}$ $(V/m)^{-2}$. The electric field of the light can be expressed as:

$$E(x, r, t) = R(r)\varphi(x, t) \exp(jk_0x - j\omega_0 t) \tag{7.9}$$

where $\varphi(x, t)$ represents the optical pulse waveform along the fiber, and $R(r)$ is the radial dependence of the electric field. It was shown that when the medium is dispersive, evolution of the waveform φ can be described in the following form [17]:

$$j\left(\frac{\partial\varphi}{\partial t} + u_0\frac{\partial\varphi}{\partial x} + v_0\varphi\right) + \frac{1}{2}u_0'\frac{\partial^2\varphi}{\partial x^2} + \frac{\alpha\omega_0 n_2}{n_0}|\varphi|^2\varphi = 0 \tag{7.10}$$

where $u_0 = \dfrac{\partial\omega}{\partial k_0}$, $u_0' = \dfrac{\partial^2\omega}{\partial k_0^2}$, $v_0 = \dfrac{\chi(\omega_0)}{n_0}\omega_0$ and α is an averaging factor that depends on the radial variation of the field. The group velocity of the light pulse is u_0, while u_0' represents the dispersion of the refractive index.

The fourth term represents the broadening of the wave form due to the dispersion, and the fifth term describes the nonlinear effect of the optical electric field. We expect that if u_0' has a positive value, then the fourth term can be compensated by the fifth term when $(\partial^2\varphi/\partial x^2)\,\varphi^{-1} < 0$ holds; that is, when the optical pulse shape is convex upward. This means that there can be a stationary pulse form in a wavelength region where $\dfrac{\partial^2\omega}{\partial k^2} > 0$, which is usually called the anomalous dispersion region. As was shown in Figure 7.3, the anomalous dispersion region of silica fiber corresponds to $\lambda > 1.3$ μm.

It was shown that there are actually stable solutions for (7.10), which are called solitons, the most fundamental of which can be described as:

$$\varphi(x, t) = E_s\text{sech}\left(\frac{t - t_0 - x/u_g}{\tau_0}\right)\exp[j(Kx - \Omega t)] \tag{7.11}$$

with

$$E_s^2 = n_0 u_0' / \partial \omega_0 n_2 u_g^2 \tau_0^2$$

$$u_g = u_0 + K u_0'$$

$$\Omega = K u_0 + \frac{1}{2} K^2 u_0' - \alpha \omega_0 n_2 E_s^2 / 2 n_0 \tag{7.12}$$

Although five parameters (E_s, τ_0, Ω, K, and u_g) appear in (7.11), only two can be chosen as independent parameters determined by the initial conditions, and the other three are then determined by (7.12). In (7.11), τ_0 means the pulse width. It should be noted that, from (7.12), the peak power P_s of the stable soliton is inversely proportional to τ_0^2 and is given by:

$$P_s = \frac{1}{2} n_0 E_s^2 \left(\frac{\varepsilon_0}{\mu_0}\right)^{1/2} S \approx \frac{10^7 S(\mu m)^2}{(\omega_0 \tau_0)^2} \left| \frac{\omega_0}{u_g} \frac{\partial u_g}{\partial \omega_0} \right| \tag{7.13}$$

Hasegawa and Tappert estimated the peak power to be $P_s \simeq 90$ mW for a pulse width $\tau_0 = 3$ ps, effective fiber cross sectional area $S = 10\ \mu\text{m}^2$, and $\frac{\omega_0}{u_g} \frac{\partial u_g}{\partial \omega_0} \simeq 5 \times 10^{-2}$. Although a high peak power is necessary for the soliton propagation, it is within the practical power range of laser diodes. Figure 7.9 shows the effect of soliton propagation over a distance of 1.8 km, simulated by Hasegawa and Tappert. While the linear pulse spreads from 3 ps to over 20 ps, there is almost no change in the waveform for the soliton.

The soliton should spread out as a result of the fiber optical loss, as expected from (7.13). However, the loss can be compensated by the Raman amplification and erbium-doped fiber amplification, so that over long distances, solitons can form at very low powers.

As was described in Section 2.5, the Raman gain band in optical fibers is widely spread over a terahertz bandwidth. Therefore, very short pulses can be amplified without deterioration of the pulse shape. Although the Raman gain of a silica fiber is not as high as that of an Er^{3+}-doped fiber, it is sufficient to compensate for the very small optical loss that occurs in a fiber.

It is an advantage that femtosecond light pulses can be transported in a fiber [18]. Figure 7.10 shows an experimental result. Light pulses as short as 80 fs with $\lambda \simeq 1.3 \sim 1.5\ \mu\text{m}$ were propagated in a 300m fiber with a Raman pump power of

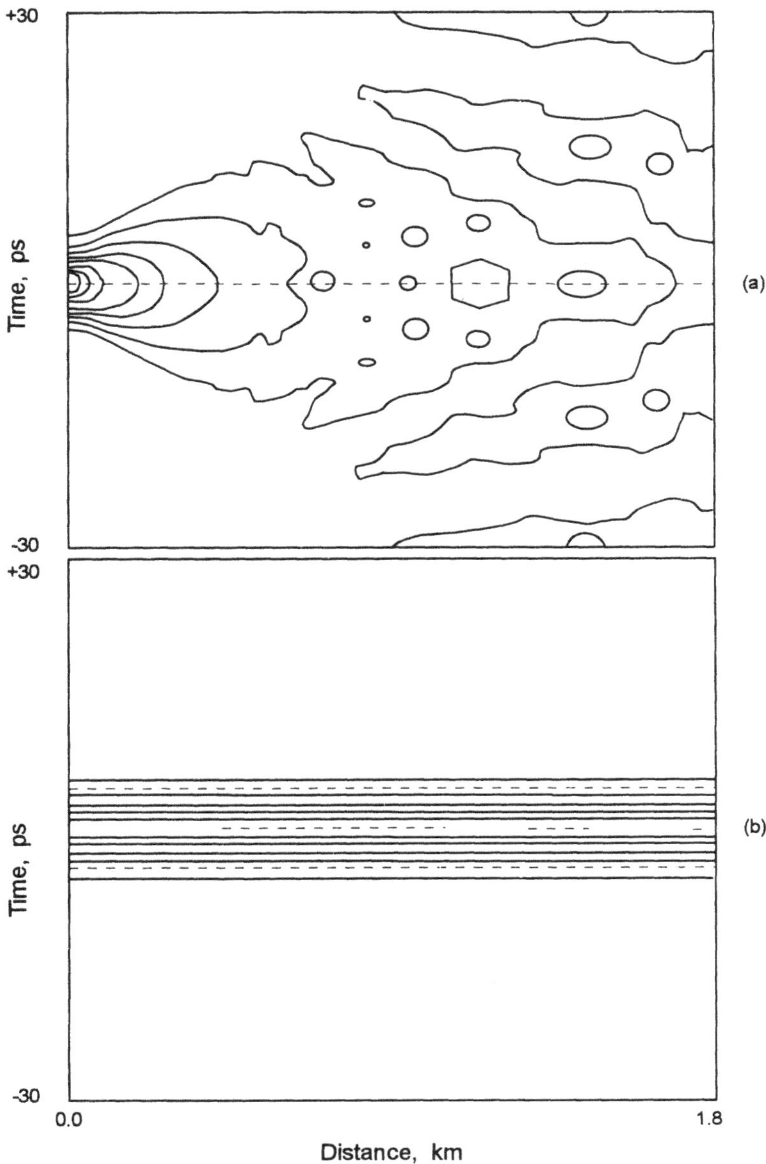

Figure 7.9 Comparison of linear (a) and stationary nonlinear (b) propagation of 3-*ps* optical pulses in glass fibers [17].

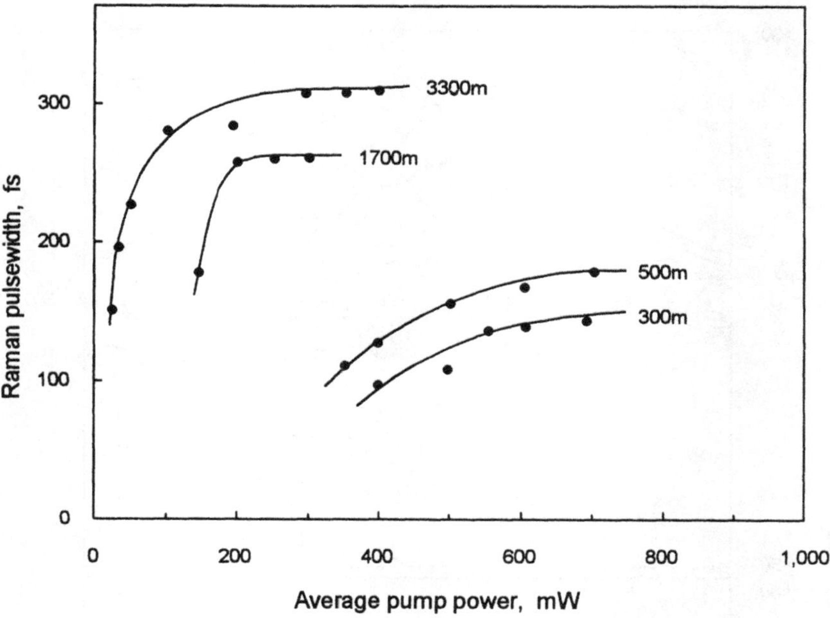

Figure 7.10 Variation of the soliton Raman pulse width with average pump power in the fiber for various fiber lengths [18].

approximately 300 mW and pump wavelength of 1.32 μm. Raman pump pulse, with width of ~100 ps and wavelength 1.32 μm from a *cw* actively mode-locked Nd YAG laser were introduced from the end of a single-mode fiber. First-Stokes Raman oscillation builds up in the early stage of the fiber to form solitons, and then Raman amplification and absorption loss balance to transmit stable solitons. The Raman amplification band extend from 1.32 to 1.54 μm, so that narrow solitons can be transported.

There are challenges for long-distance soliton propagation by means of Raman amplification. As illustrated in Figure 7.11, Raman amplification is repeated with an interval of L [19,20]. It is expected that few picosecond pulses can be propagated over distances >1000 km [20]. However, it is a disadvantage that the amplification period L must not be larger than several tens of kilometers, so a large number of repetitions are necessary for long-distance propagation.

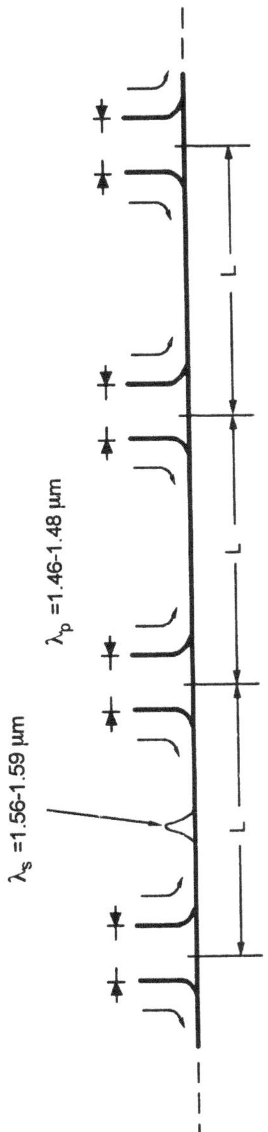

Figure 7.11 Illustration of periodically Raman-pumped long-distance soliton propagation [20].

7.2 LASER DIODE AMPLIFIER

Although the Er^{3+}-doped optical fiber amplifier is very suitable for present-day optical communication systems, it has a disadvantage in that the wavelength of amplification is limited to approximately 1.53 μm and/or 1.55 μm. To fully utilize the tremendously wide range of the light frequency or wavelength, other kinds of light amplifiers are of interest. The laser diode amplifier, which has been studied from as long ago as the 1960s, can be used at a wide variety of wavelengths, although some difficulties must be overcome for practical applications.

One of the earliest amplifier experiments was made by Nishizawa, as shown in Figure 7.12 [21]. Two laser diodes were placed face to face, separated by a small

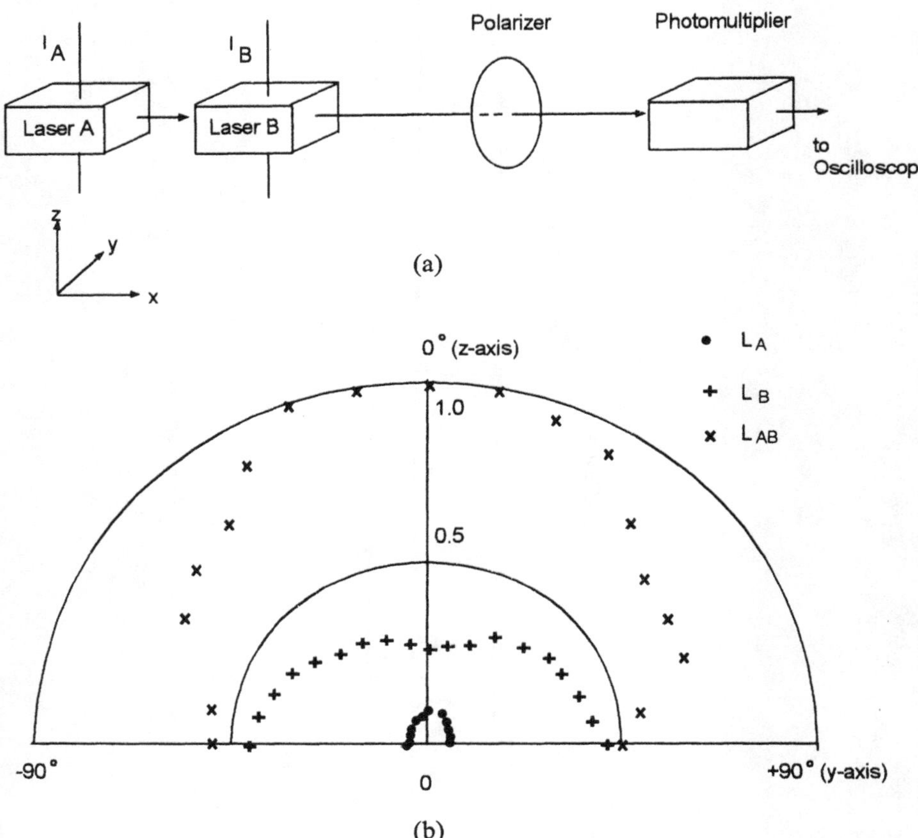

Figure 7.12 (a) Schematic model of amplification experiment (b) polarization in the interaction of laser light (junction planes of both lasers are in *X-Y* plane) [21].

distance by using a special manipulator controlled by an electromagnet. It was confirmed that the polarization of the laser diode A is conserved after amplification by the laser diode B.

Even now, the coupling of the incident light beam to the amplifier is one of the essential problems because the thickness of the active layer is only on the order of 1,000Å. There is usually approximately 10 dB of coupling loss so that, even for an amplifier with a gain as high as 20 dB, the net gain reduces to about 10 dB. The light reflected from the amplifier facets, as well as from components of the output optical circuit getting into the amplifier, must be carefully eliminated because it is strongly amplified in the laser diode, resulting in instabilities and noise. Therefore, both end-faces of an amplifier are antireflection coated, and an optical isolator is placed at the output of the amplifier.

In comparison to the Er^{3+}-doped fiber amplifier, the laser diode amplifier characteristics are essentially different in that the spontaneous lifetime is very small, less than 1 ns. Moreover, if the coherent light field is confined in a very narrow region, as occurs in the laser diode oscillator, the lifetime can be greatly reduced according to the following equation [22]:

$$\frac{1}{\tau} = \frac{1}{\tau_{sp}} + A|E|^2 \qquad (7.14)$$

When there are two optical fields, with frequencies ω_A and ω_B, present in the laser diode amplifier, a nonlinear interaction between two fields occurs because the population inversion is modulated at the difference-frequency $\omega_D = \omega_A - \omega_B$. As shown in Figure 7.13, when the frequency from the diode A gets into the regenerative amplification band width of the amplifier B, a new line appears at a frequency $2\omega_B - \omega_A$ [23]. This fact means that the light beam with ω_B is modulated at the difference-frequency ω_D. The modulation was found to be observable up to $\omega_D \simeq$ 90 GHz, corresponding to 1-ps response time, as shown in Figure 7.13(b). This high-speed modulation is due to the shortened lifetime according to (7.14). Although a similar effect will occur in other kinds of lasers and amplifiers, it is of particular importance in the laser diode because the optical field confined in a narrow active region can easily build up to a high value.

This nonlinear interaction effect is very attractive for high-speed optical switching as a basis for the optical computer [24]. However, it also causes serious distortion of the pulse waveform and introduces crosstalk between different channels of a wavelength-division-multiplexed signal.

To reduce the nonlinear effect, the laser diode should not be operated in the saturation region where the gain is strongly reduced. This is the same as for the Er^{3+}-doped fiber amplifier.

Detailed amplifier characteristics of double-heterostructure laser diode amplifiers are discussed in [25]. To increase the saturation output power, the laser diode current injection level must be considerably higher than that of laser diode oscillators. The multiquantum well (MQW) laser diode, illustrated in Figure 7.14, is thought to be superior to conventional laser diodes because it requires a relatively low current for operation at a high injection level. The MQW amplifier also has low noise figure characteristics because a highly inverted state is easily attained [26,27]. Even if an optical field builds up to increase the stimulated emission transition for the excess (or inverted) carriers within a well, they are rapidly supplied from the higher levels in the well, as well as from the barrier regions, so that the population inversion is not reduced so much.

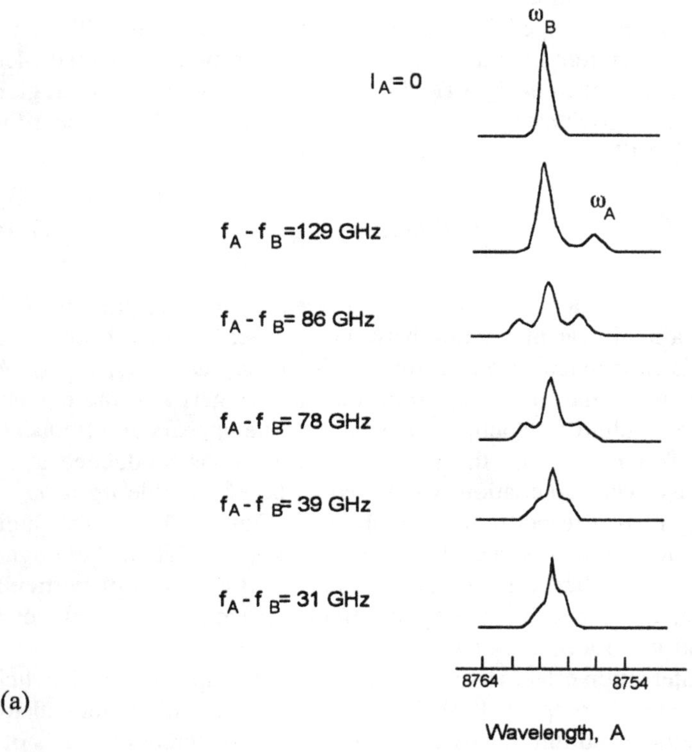

Figure 7.13 (a) The change of the output spectra of a laser diode B as the lasing grequency of a laser diode A is varied. The length of B is 70 μm, which corresponds to the fundamental mode frequency of 506 GHz. (b) The frequency range of modulation for l_B = 90 μm and l_B = 70 μm. δ is the difference between the lowest and the highest frequency in the output of B. $P = 1$ indicates that injection-induced modulation does not occur, and $P = 2$ indicates the occurrence of modulation [23].

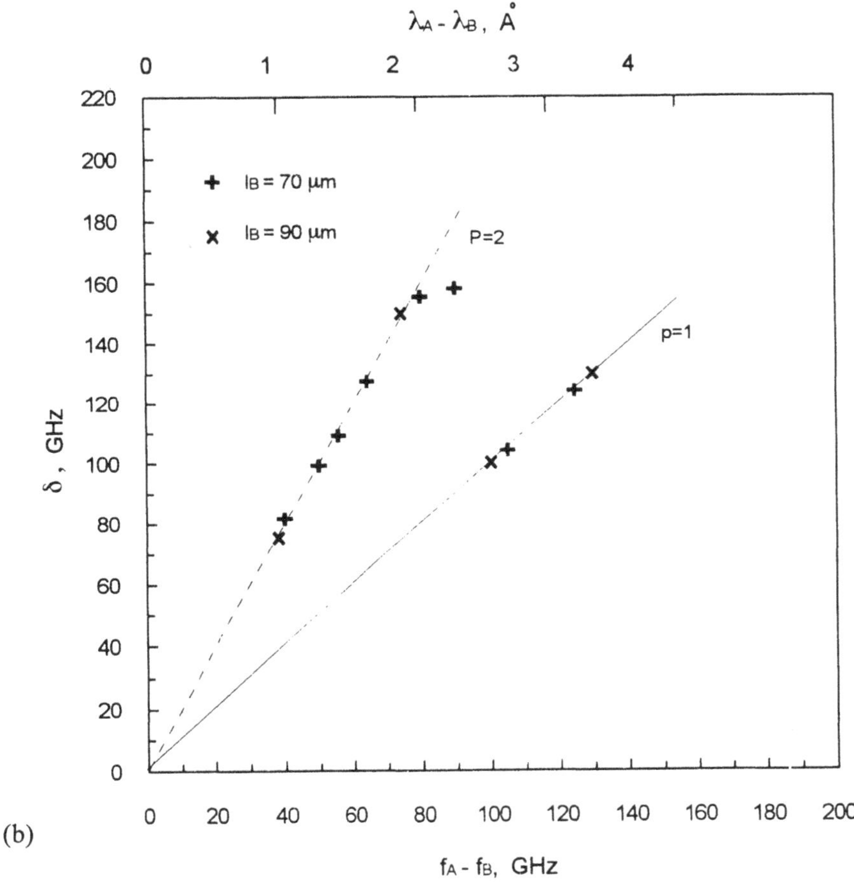

(b)

Figure 7.13 (Continued)

Figure 7.15 shows an example of the amplifier characteristics of an MQW laser diode [28]. The disadvantages of the MQW laser diode are strong polarization effects and lower maximum gain. Because of the anisotropic band structure in the quantum well, the gain coefficient is higher for the TE mode than for the TM mode. On the other hand, the lower maximum gain is due to the smaller optical confinement factor for the MQW structure. The polarization dependence of the gain can be eliminated in an InGaAs MQW amplifier that comprises alternatively strained tensile and compressive quantum wells to balance the polarization-dependent gain of each type of well [25].

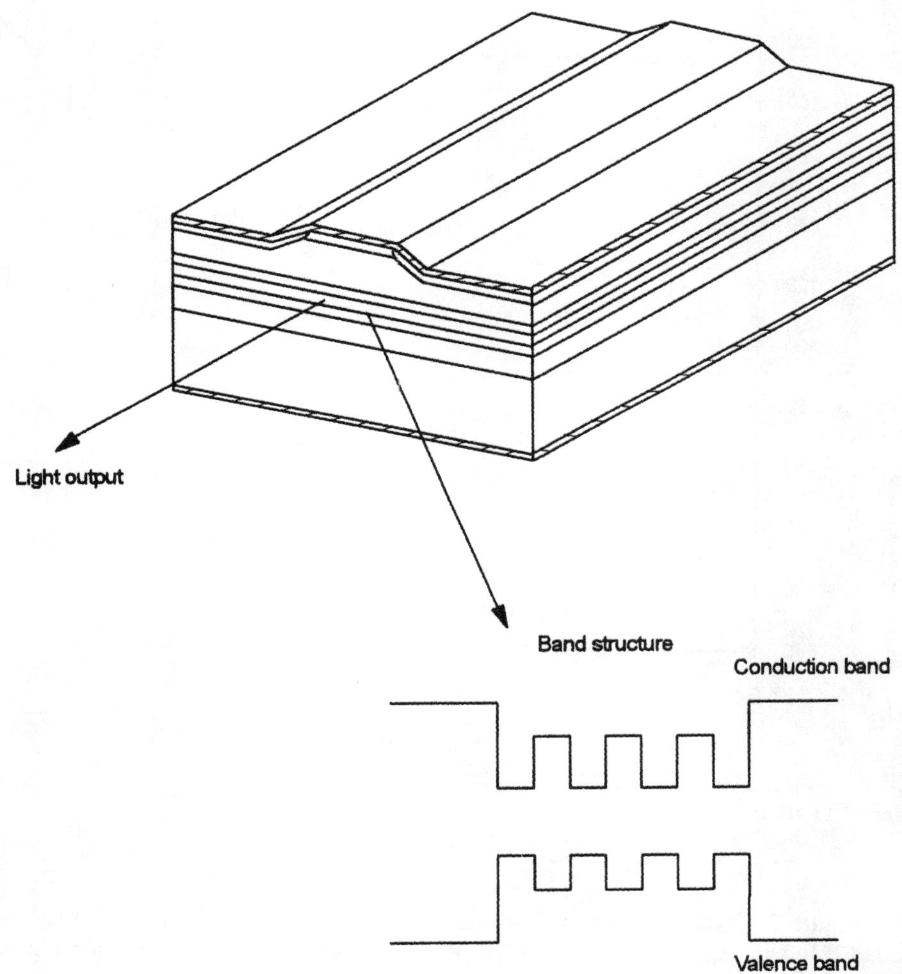

Figure 7.14 Structure of an MQW laser diode amplifier.

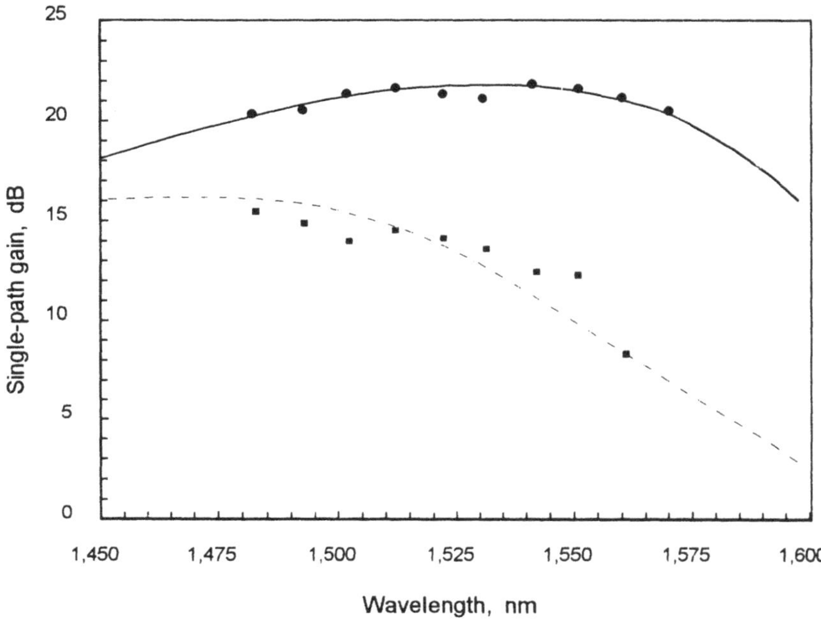

Figure 7.15 InGaAs MQW laser diode amplifier. Measured and calculated results for the single-pass gain versus wavelength at an injection current of 225 mA (TE = ●, TM = ■) [27].

The gain bandwidth of a laser diode amplifier extends to over 1 THz. Therefore, it can be used for inline repeaters as well as postamplifiers in very wideband optical communications. However, it should be remembered that spontaneous emission noise increases in proportion to the bandwidth Δf, as discussed in Section 7.1. Optical filters are used to reduce the spontaneous emission noise reaching the detector, as in the case of Er^{3+}-doped optical fiber.

7.3 DEMODULATION BY THE RAMAN LASER IN WIDEBAND OPTICAL COMMUNICATION

7.3.1 Demodulation Experiment

As was mentioned in Chapter 1, there are two different approaches for future optical communication systems: wideband optical communication based on frequency division and high-speed optical communication based on time division.

Terahertz-bandwidth optical communications based on frequency division is aimed at fully utilizing the information capacity of light. It should be remembered that the response frequency present limit of the fastest light detector, pin photodiodes as well as avalanche photodiodes and Schottky diodes, is less than 100 GHz in GaAs devices, so they cannot be applied to terahertz-modulated light. Thus, the semiconductor Raman laser amplifier will have a possible role in a wideband optical communication as a heterodyne-type demodulator [29]. Other approaches are possible to demultiplex high data rate signals (e.g., nonlinear switching).

Figure 7.16 illustrates the principle of a terahertz-bandwidth optical communication system that utilizes the semiconductor Raman laser amplifier [30,31]. A wideband-modulated light wave can be thought of as being composed of closely separated frequency components ω_{s1}, ω_{s2}. ... ω_{si}. ... , with a frequency interval of $\Delta\omega$. The semiconductor Raman laser amplifier picks up a desired frequency component, ω_{si}, by tuning the frequency ω_L of the pump laser diode to fulfil the condition $\omega_L = \omega_{si} + \omega_{ph}$. The optical phonon frequency of GaP crystals, ω_{ph}, has the fixed value of 12 THz. The laser diode frequency ω_L can be tuned over a few terahertz by changing the current only.

In other words, the semiconductor Raman laser acts as a frequency-selective light amplifier. The frequency bandwidth $\dfrac{\Delta\omega}{2\pi}$ of the amplifier is approximately 30 GHz, which corresponds to the damping constant of the LO phonons in GaP. If the amplifier has a partially reflective resonator structure, the bandwidth $\Delta\omega$ can be designed to give any other smaller value, with an increased gain, as a result of regenerative amplification. In any case, a 1 to 10 G bit/s optical pulse train with a center frequency ω_{si} will drop into one channel. Signal frequency selection can be made very fast because it is accomplished by adjusting the current supplied to the pump laser diode.

Although we have presumed a terahertz-modulated light signal, we can instead consider the case that multiple light signals with closely separated frequencies from a number of signal source laser diodes come into a demodulator. The semiconductor Raman laser amplifier can pick up any one from those signals. Therefore, the proposed system can be constructed as a simple extension of the present optical communication systems.

We have carried out an experiment showing frequency-selective light amplification, which is a demonstration of the proposed demodulation method [32].

Figure 7.17 illustrates the experimental system. We have used two wavelength tunable Ti-sapphire lasers, one for a pump laser and the other for the signal light source. The pump Ti-sapphire laser is a conventional reflector resonator type equipped with thin and thick etalons for frequency stabilization, but it operates in two or three longitudinal modes with a frequency separation of 200 MHz. The signal Ti-sapphire laser is a ring laser that operates in a single longitudinal mode. It is also equipped with a thick etalon. Therefore, the light frequency can be varied step-by-step in intervals of 20 GHz but not tuned continuously.

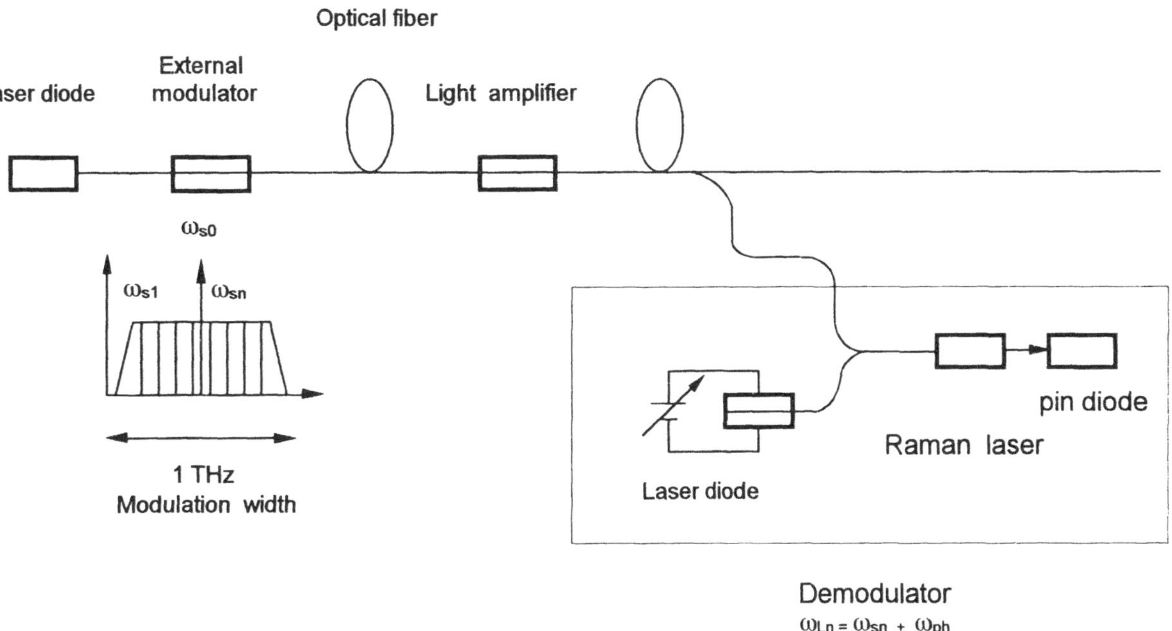

Figure 7.16 Terahertz-band optical communication system with a Raman laser demodulator.

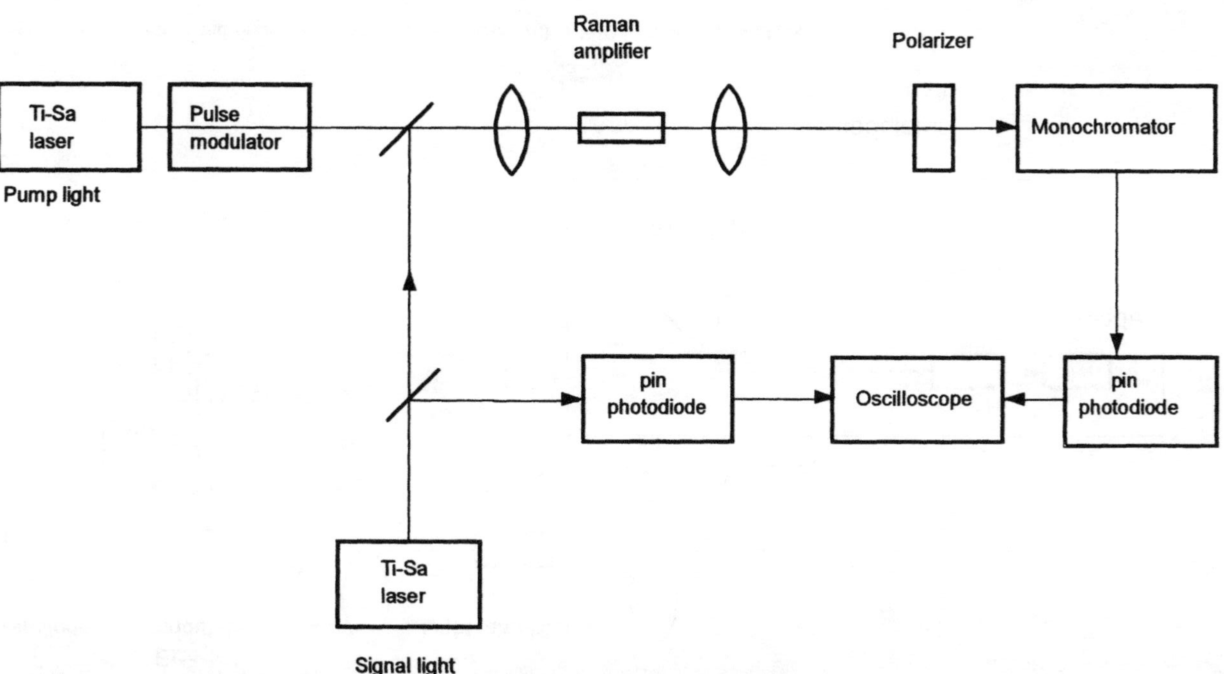

Figure 7.17 Experimental system for a frequency-selective single-pass Raman amplification.

The waveguide of the semiconductor Raman laser amplifier is the same as that of the Raman laser oscillator described in Section 4.3, except that both the input and output surfaces are antireflection coated with a transmittance $T \geq 99.5\%$ in the wavelength region of 820 ~ 940 nm.

In this experiment, the GaP core is 6-μm wide and 4.3-μm thick. The pump beam diameter at the input surface is approximately 4 μm, and the transmitted pump light intensity is 320 mW so that the internal pump power density is $P_L \simeq 1.3 \times 10^6$ W/cm^2. The frequency of the signal laser can be of any value, provided both the pump and signal wavelengths are in the transmission region. It has been fixed at one arbitrary frequency ω_i = 324 THz (i.e., the wavelength λ_i = 925 nm), and the frequency of the pump laser has been varied around the supposed center frequency ω_L = 336 THz (λ_{LO} = 892 nm). Actually, ω_L cannot be continuously tuned. Rather, it can be tuned only in steps of 20 GHz, which is the free spectral range of the internal thick etalon of the pump laser.

The pump and signal light waves propagate in the same forward direction in fundamental transverse modes. The pump beam going out from the Raman amplifier is removed by a polarizer as well as a monochromator, and the signal output is detected by an Si pin photodiode. The signal laser beam is in the *cw* mode of operation, and the input signal power level is 15 mW. The pump beam is pulse modulated with a pulse width of 2.5 μs. Therefore, only the amplified component can be detected as the pulse signal.

However, we have made differential type detection using a reference pin photodiode to reduce the noise level due to the *cw* signal light background, as illustrated in Figure 7.17. The result of the measurement is shown in Figure 7.18.

No output signal is detected when the difference between the pump frequency and the center frequency, which gives the maximum gain, is greater than 40 GHz. Thus, the amplification bandwidth can be estimated to be $\Delta f \simeq 30$ GHz, as indicated by the curve in Figure 7.18. The actual gain at the center frequency was approximately 2% to 2.5%, when the length of the amplifier was 4 μm and the pump power density $P_0 = 1.3 \times 10^6$ W/cm^2. Therefore, the gain per cm per 10^6 W/cm^2 pump power density is given by:

$$g_0 \approx 0.038 \sim 0.047/(cm) \cdot (10^6 \text{ W/cm}^{-2}) \qquad (7.15)$$

at a pump wavelength λ_p = 892 nm.

Although this value is apparently small, it should be noted that the gain thus obtained is for the forward Raman scattering only. For the Raman laser having high-reflection mirrors, both the forward and the backward scattering contribute to the gain. For the longitudinal optical phonon mode, there is little frequency dispersion, so we can assume that the gains for the forward and backward scattering are nearly the same. Then, the round-trip gain for a laser is 4 times g_0, not 2 times.

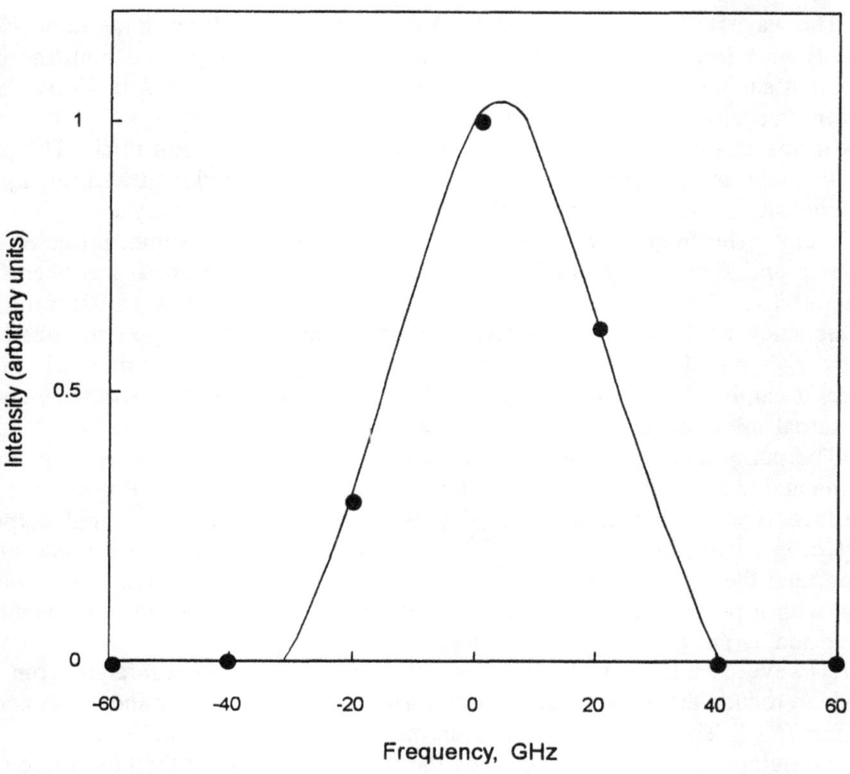

Figure 7.18 Raman-amplified output as a function of the pump frequency offset.

The situation is the same for the amplifier equipped with a high-reflection mirror at the back-end surfaces of the waveguide, which is shown in Figure 7.19 in Section 7.3.2. It should also be noted that the pump power density and the signal power density can be enhanced if the input side is made partially reflective.

7.3.2 Regenerative and Frequency-Selective Light Amplification

Experimental evidence of the regenerative amplification was described in Section 3.4. Let us consider the Raman amplifier of the construction shown in Figure 7.19, with the input reflector having amplitude reflectances at the pump wavelength and the Stokes wavelength r_{pump} and r, respectively, and with the end reflector having an amplitude reflectance $r' = -1$ (i.e., 100% reflectance).

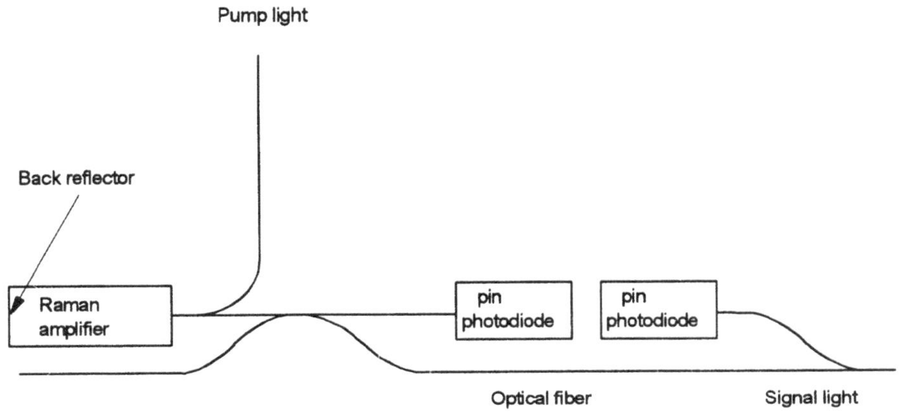

Figure 7.19 Raman laser amplifier with a back reflector.

For the pump light, it was shown in (5.18) in Section 5.3.2 that the internal pump power in the waveguide is enhanced by a factor k given by:

$$k = \frac{1 + r_{pump}}{1 - r_{pump}} \tag{7.16}$$

On the other hand, the partial reflection at the input reflector results in regenerative amplification, which causes gain enhancement and narrowing of the amplifier bandwidth. The Stokes field output at the input surface of the amplifier is given as a sum of reflected fields as follows:

$$\begin{aligned} E_{out} &= E_{in}\{r + t\bar{t}r' \exp i\delta \exp 2\gamma l[1 + (-rr') \exp i\delta \exp i\delta \exp 2\gamma l \\ &\quad + (-rr')^2(\exp i\delta \exp 2\gamma l)^2 + ..]\} \\ &= E_{in}\left\{\frac{t\bar{t} \exp i\delta \exp 2\gamma l}{1 + rr' \exp i\delta \exp 2\gamma l} + r\right\} \end{aligned}$$

with

$$\delta = \frac{2\pi}{\lambda} 2nl, \quad \gamma = \frac{g}{2}, \quad t\bar{t} = 1 - r^2 \tag{7.17}$$

where t and $\bar{t}$ are the amplitude transmittance and its conjugate, γ and g are the amplitude and power amplification factors, respectively, and δ is the phase shift for a round trip. Thus, the power gain for the regenerative amplification, G_{reg}, is given by:

$$G_{reg} = \left|\frac{E_{out}}{E_{in}}\right|^2 = \left|\frac{r + r' \exp i\delta \exp 2\gamma l}{1 + r_s r' \exp i\delta \exp 2\gamma l}\right|^2$$
$$= \frac{r^2 + (r' \exp 2\gamma l)^2 + 2rr' \exp 2\gamma l \cos\delta}{1 + 2rr' \exp 2\gamma l \cos\delta + (rr' \exp 2\gamma l)^2} \tag{7.18}$$

When $r' = -1$, the equation becomes:

$$G_{reg} = \left(\frac{r - \sqrt{G}}{1 - r\sqrt{G}}\right)^2 \frac{1 + \dfrac{4r\sqrt{G}}{(r - \sqrt{G})^2} \sin^2 \dfrac{\delta}{2}}{1 + \dfrac{4r\sqrt{G}}{(1 - r\sqrt{G})^2} \sin^2 \dfrac{\delta}{2}} = G_M \frac{1 + \dfrac{F}{G_M} \sin^2 \dfrac{\delta}{2}}{1 + F\sin^2 \dfrac{\delta}{2}} \tag{7.19}$$

with

$$G_M = \left(\frac{r - \sqrt{G}}{1 - r\sqrt{G}}\right)^2, \quad F = \frac{4r\sqrt{G}}{(1 - r\sqrt{G})^2}, \quad \text{and} \quad \sqrt{G} = \exp 2\gamma l = \exp gl$$

where G means a round-trip power gain, G_M is the maximum power gain at $\delta = 0$, and F corresponds to a finesse. When G_M is large, the 3-dB amplification bandwidth Δf is given by:

$$\Delta f \fallingdotseq \Delta f_0 \frac{2}{\pi\sqrt{F}} \tag{7.20}$$

with

$$\Delta f_0 = \frac{c}{2nl}$$

where Δf_0 is the frequency spacing between longitudinal modes of the resonator, c is the light velocity in vacuum, and n is the refractive index.

Figure 7.20 presents results of G_M and Δf calculation versus G for some different values of the amplitude reflectance r of the input reflector. For example, if we choose the input power reflectance $r^2 = 0.25$ ($r = 0.5$), a maximum gain of $G_M = 15$ dB with an amplification bandwidth of 1 GHz can be obtained when the round-trip gain G is 4 db. The above values of the gain and bandwidth are sufficient for a preamplifier that is to be used in connection with a pin photodiode detector.

By employing the tapered structure discussed in Section 5.5, we can obtain an amplifier with a low pump power. As is given in Table 7.1, it is expected that when the waveguide cross section is 1 μm by 1 μm and the amplifier length is 4 mm, a 15-dB gain with a 1-GHz bandwidth or a 4-dB gain with a 30-GHz bandwidth can be obtained at a pump power level of 30 mW. The tapered waveguide amplifier will cause almost no insertion losses for the pump beam and the Stokes beam.

Noise in the Raman amplifier is given by the same form as in (7.6) in Section 7.1:

$$\sigma_{out}^2 = G_M \langle n_{in} \rangle + (G_M - 1)n_{sp}m_t\Delta f + 2G_M(G_M - 1)n_{sp}\langle n_{in} \rangle + (G_M - 1)^2 n_{sp}m_t\Delta f,$$

but the spontaneous emission factor, n_{sp}, for the Stokes radiation is given by:

$$n_{sp} = \frac{N_1}{N_1 - N_2} = \frac{1}{1 - \exp\left(\frac{\hbar\omega_{ph}}{kT}\right)} \tag{7.21}$$

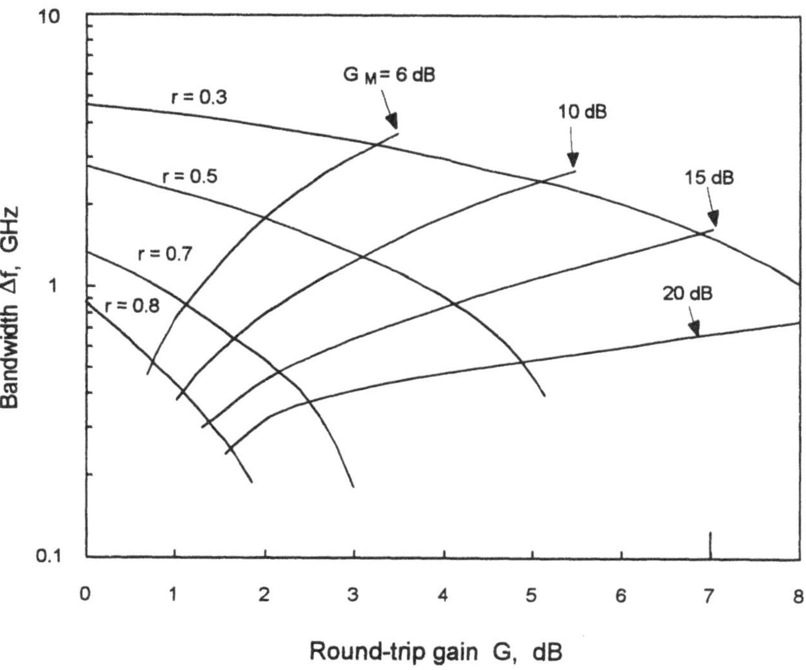

Figure 7.20 Gain and bandwidth of regenerative amplification in a GaP Raman laser amplifier.

Table 7.1
Expected Gain and Bandwidth of a Thin Waveguide Raman Laser Amplifier With a Back Reflector

Items	*Parameters*	*Gain*	*Bandwidth*
Length	4 mm		
Core size	1 × 1 μm		
Pump power	30 mW		
Enhancement factor	4		
Traveling wave			
$R_{in} = 0$		4 dB	30 GHz
Regenerative			
$R_{in} = 0.25$		15 dB	1 GHz

where N_1 and N_2 are the numbers of molecules in the ground vibrational state and the excited vibrational state, respectively. At room temperature, n_{sp} is as small as 1.16 for the longitudinal optical phonon in GaP.

The most significant feature of the semiconductor Raman laser amplifier is that the bandwidth Δf contributing to the ASE noise can be made as narrow as required for one-channel signal bandwidth, without using a filter. That is, the tunability of a system with a semiconductor Raman laser is not restricted by the filter, which is used with other amplifiers.

REFERENCES

[1] F. Inaba, "Solid State Laser" (in Japanese), *Quantum Electronics*, ed. K. Ishiguro, K. Shimoda and T. Yajima, Asakura Book Co., Tokyo, 1965, p. 17.

[2] E. Snitzer, "Proposed Fiber Cavities for Optical Masers," *J. Appl. Phys.*, Vol. 32, 1961, pp. 36–39.

[3] E. Snitzer and C.G. Young, "Glass Lasers," *Lasers 2*, ed. A.K. Levine, New York: Marcel Dekker Inc., 1968, pp. 191–256.

[4] S.B. Poole, D.N. Payne, R.J. Mears, M.E. Fermann and R.I. Laming, "Fabrication and Characterization of Low-Loss Optical Fibers Containing Rare-Earth Ions," *IEEE J. Lightwave Technol.*, Vol. LT4, 1986, pp. 870–876.

[5] N. Inagaki, T. Edahiro and M. Nakahara, "Recent Progress in VAD Fiber Fabrication Process," *Jpn. J. Appl. Phys.*, Vol. 20 Supp. 20–1, 1981, pp. 175–180.

[6] A. Sugimura, K. Daikoku, N. Imoto and T. Miya, "Wavelength Dispersion Characteristics of Single-Mode Fibers in Low-Loss Region," *IEEE J. Quantum Electronics*, Vol. QE16, 1980, pp. 215–225.

[7] Y. Ohishi, T. Kanamori, T. Nishi, S. Takahashi and E. Snitzer, "Concentration Effect on Gain of Pr^{3+}-Doped Fluoride Fiber for 1.3 μm Amplification," *IEEE Photon. Technol. Lett.*, Vol. 4, 1992, pp. 1338–1340.

[8] M. Shimizu, T. Kanamori, J. Temmyo, M. Wada, M. Yamada, Y. Terunuma, Y. Ohishi and S. Sudo, "28.3 dB Gain 1.3 μm-Band Pr-Doped Fluoride Fiber Amplifier Module Pumped by 1.017 μm InGaAs-LD's," *IEEE Photon. Technol. Lett.*, Vol. 5, 1993, pp. 654–657.
[9] E. Desurvire and J.R. Simpson, "Amplification of Spontaneous Emission in Erbium-Doped Single-Mode Fibers," *J. Lightwave Technol.*, Vol. 7, 1989, pp. 835–845.
[10] E. Desurvire, J.L. Zyskind, and J.R. Simpson, "Spectral Gain Hole-Burning at 1.53 μm in Erbium-Doped Fiber Amplifiers," *IEEE Photon. Technol. Lett.*, Vol. 2, 1990, pp. 246–248.
[11] J.L. Zyskind and J.R. Simpson, "Determination of Homogeneous Linewidth by Spectral Gain Hole-Burning in an Erbirm-Doped Fiber Amplifier with GeO_2 : SiO_2 Core," *IEEE Photon. Technol. Lett.*, Vol. 2, 1990, pp. 869–871.
[12] H. Ishio, S. Nakagawa, M. Nakazawa, K. Aida and K. Hagimoto, *Light Amplifiers and Their Applications* (in Japanese), OHMSHA LTD, 1992.
[13] H. Masuda and A. Takada, "High Gain Two-Stage Amplification With Erbirm-Doped Fiber Amplifier," *Electron. Lett.*, Vol. 26, 1990, pp. 661–662.
[14] J.P. Gordon and H.A. Haus, "Random Walk of Coherently Amplified Solitons in Optical Fiber Transmission," *Opt. Lett.*, Vol. 11, 1986, pp. 665–667.
[15] R.I. Laming, M.N. Zervas, and D.N. Payne, "Erbium-Doped Fiber Amplifier With 54 dB Gain and 3.1 dB Noise Figure," *IEEE Photon. Technol. Lett.*, Vol. 4, 1992, pp. 1345–1347.
[16] S.M. Hwang and A.E. Willner, "Guidelines for Optimizing System Performance for 20 WDM Channels Propagating Through a Cascade of EDFA's," *IEEE Photon. Technol. Lett.*, Vol. 5, 1993, pp. 1190–1193.
[17] A. Hasegawa and F. Tappert, "Transmission of Stationary Nonlinear Optical Pulses in Dispersive Dielectric Fibers 1. Anomalous Dispersion," *Appl. Phys. Lett.*, Vol. 23, 1973, pp. 142–144.
[18] A.S. Gouveia, A.S.L. Gomes and J.R. Taylor, "Femtosecond Soliton Raman Generation," *IEEE J. Quantum Electron.*, Vol. 34, 1988, pp. 332–340.
[19] A. Hasegawa, "Numerical Study of Optical Soliton Transmission Amplified Periodically by the Stimulated Raman Process," *Appl. Optics*, Vol. 23, 1984, pp. 3302–309.
[20] L.F. Mollenauer, J.P. Gordon and M.N. Islam, "Soliton Propagation in Long Fibers With Periodically Compensated Loss," *IEEE J. Quantum Electron.*, Vol. QE-22, 1986, pp. 157–173.
[21] J. Nishizawa, M. Takusagawa, S. Ohsaka, Y. Goto and M. Suzuki, "Conservation of Polarization in GaAs Junction Laser," *Jpn. J. Appl. Phys.*, Vol. 11, 1972, pp. 419–420.
[22] J. Nishizawa, "Recombination Lifetime in a Semiconductor Laser Diode," *IEEE J. Quantum Electron.*, Vol. QE4, 1968, pp. 143–147.
[23] J. Nishizawa and K. Ishida, "Injection-Induced Modulation of Laser Light by the Interaction of Laser Diodes," *IEEE J. Quantum Electron.*, Vol. QE11, 1975, pp. 515–519.
[24] J. Nishizawa, "Interaction of Light Beams in Semiconductor Laser Diode," *Jpn. J. Appl. Phys.*, Supp. 43, 1974, pp. 89–94.
[25] T. Mukai, Y. Yamamoto and T. Kimura, "Optical Amplification by Semiconductor Lasers," *Semiconductor and Semimetals*, Vol. 22, Part E, Chap. 3, 1985, pp. 265–319.
[26] T. Saitoh, Y. Suzuki and H. Tanaka,"Low Noise Characteristics of a GaAs-AlGaAs Multi-Quantum-Well Semiconductor Laser Amplifier," *IEEE Photon. Technol. Lett.*, Vol. 2, 1990, pp. 794–796.
[27] K.S. Jepsen, B. Mikkelsen, J.H. Povlsen, M. Yamaguchi and K.E. Stubkjaer, "Wavelength Dependence of Noise Figure in InGaAs/InGaAsP Multiple-Quantum-Well Laser Amplifier," *IEEE Photon. Technol. Lett.*, Vol. 4, 1992, pp. 550–553.
[28] M.A. Newkirk, B.I. Miller, U. Koren, M.G. Young, M. Chien, R.M. Jopson and C.A. Burns, "1.5 μm Multiquantum-Well Semiconductor Optical Amplifier With Tensile and Compressively Strained Wells for Polarization-Independent Gain," *IEEE Photon. Technol. Lett.*, Vol. 4, 1993, pp. 406–408.
[29] J. Nishizawa and K. Suto, "Lightwave Demodulator," Japanese Patent 1605283, Application 1981.

[30] K. Suto and J. Nishizawa, "Characteristics of the Epitaxial Semiconductor Raman Laser," *IEE PROC.*, Vol. 133, Pt. J, 1986, pp. 259–263.

[31] K. Suto, S. Ogasawana, T. Kimura and J. Nishizawa, "Semiconductor Raman Laser as a Tool for Wideband Optical Communications," *IEE PROC.*, Vol. 137, Pt. J, 1990, pp. 43–48.

[32] K. Suto, "Semiconductor Raman Laser for Optical Communication," *ERATO Summary Reports of NISHIZAWA Terahertz Project*, 1992, pp. 38–46.

Chapter 8
Future of the Semiconductor Raman Laser

8.1 HIGH-COHERENCY NATURE OF THE RAMAN OSCILLATOR

The linewidth of a laser is primarily affected by the spontaneous emission. Following the discussion for the laser diode linewidth by Henry [1], the spontaneously emitted photons cause a frequency broadening given by:

$$\Delta f = \frac{v_g^2 h\nu g n_{sp} \alpha_m (1 + \alpha^2)}{8\pi p_o} \tag{8.1}$$

with $g = \alpha_L + \alpha_m$ and $\alpha_m = -L^{-1}\ln R_m$, where v_g is the group velocity of photons, g the gain, α_L the waveguide loss, α_m the resonator mirror loss, L the laser length, R_m the mirror reflectance, n_{sp} the spontaneous emission factor, and P_0 the output power. The parameter α is a linewidth enhancement factor representing the refractive index change caused by the gain. In the case of laser diodes, α is as large as 2 to 10, but for the Raman laser, α can be neglected. The linewidth of laser diodes predicted by the above equation is on the order of 10 MHz.

For the estimation of the linewidth for the semiconductor Raman laser, we assume the following typical parameter values: L = 5 mm, R_m = 0.98, a_L = 0.05 cm^{-1} (this value corresponds to the carrier concentration of $n = 1 \times 10^{16}$ cm^{-3}), $n_{sp} \simeq 1$, and P_o = 1 mW. Then, we have Δf = 2 Hz. Thus, the expected value of the linewidth is almost 10^{-6} times that of the laser diode linewidth. In other words, the high resonator Q of the Raman laser results in a narrow linewidth. Although the laser diode linewidth can be made narrow with the aid of external resonators, the Raman laser has a simpler structure, and the operation will be more stable.

The linewidth is, of course, not solely determined by the spontaneous emission. In the limit of narrow linewidth, another source of linewidth broadening is the change in the resonance frequency via thermal and acoustic fluctuation. To reduce the thermal effect, the internal absorption (e.g., free-carrier absorption) must be reduced to as small as possible.

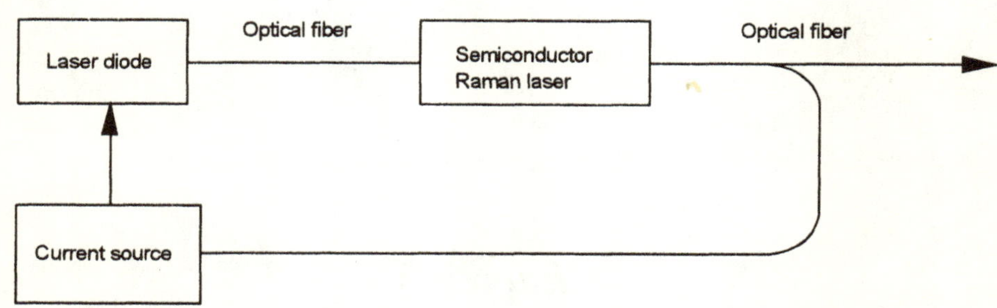

Figure 8.1 Pump laser diode locked to one of the resonance frequencies of the Raman laser waveguide.

There will be a number of important applications relying on the narrow linewidth, or in other words, high coherency of the semiconductor Raman laser. Even for the signal light source in the wideband optical communication, the Raman laser will be more suitable than the laser diode if the external modulation method is adopted. In any case, the Raman laser must be pumped by a laser diode. However, the tolerance for the frequency stability as well as the linewidth of the laser diode can be as large as the Raman gain band. As is illustrated in Figure 8.1, it will be preferable to stabilize a pump laser diode frequency by locking it to the resonance frequency of a Raman laser waveguide. This will cause another benefit, which is that the internal pump field increases by a factor $k = \dfrac{1 + r_{pump}}{1 - r_{pump}}$ as discussed in chapter 5.3.2, where r_{pump} is the amplitude reflectance for the pump light at the input surface of the Raman laser.

8.2 LIGHTWAVE MIXING IN THE RAMAN WAVEGUIDE

In the past experiments on the difference-frequency lightwave mixing, efficiencies were extremely low, only on the order of 10^{-6}. This was because phase matched long-length interaction was impossible. If we can apply the waveguide technique to the difference-frequency mixing, the efficiency will be greatly increased, and we will have convenient solid-state coherent light sources in the wide spectral range bridging between the lightwave and the microwave frequencies.

Chapter 3 pointed out that the perfect phase matching for the difference-frequency mixing in collinear configuration can be realized in GaP at a frequency region below the transverse optical phonon frequency (ω_{TO} = 11 THz) [2]. From this fact, if the light waves from the two laser diodes with frequencies ω_{L1} and ω_{L2} are introduced into a GaP—$Al_xGa_{1-x}P$ waveguide, efficient difference-frequency wave generation is expected [3]. Essentially, this effect needs no accompanyment of Raman laser oscillation. The efficiency will, however, increase more if the Raman laser oscillation occurs. This is because the internal Stokes power density can be larger than the pump power density when the resonator mirror reflectances are sufficiently high as discussed in Chapter 2.4. Also, the frequency of the generated wave will be stable because the changes in the mode frequencies on response to the change in the temperature are nearly the same for the two Stokes waves in a Raman resonator.

Figure 8.2 illustrates the method of difference-frequency wave generation under the Raman laser oscillation condition.

The Raman laser is pumped by two laser diodes with frequency ω_{L1} and ω_{L2}. It was shown that the Raman gain in such a case is given by [3]:

$$g = \frac{\alpha_{xyz}^2}{\varepsilon_0 cMN} \frac{1}{\omega_{LO}\Gamma} \left\{ \frac{\omega_{s1}}{n_{s1}} |E_{L1}|^2 + \frac{\omega_{s2}}{n_{s2}} |E_{L2}|^2 \right\} \tag{8.2}$$

In other words, the gain is given by $g = \beta_1 P_1 + \beta_1 P_1$, where β_1 and β_2 are the Raman gains for a unit of pump power, and P_1 and P_2 are the pump power densities at ω_{L1} and ω_{L2}, respectively. Then, it can be said that when one of the pump lasers has a sufficient high-power output for Raman oscillation, the other need not.

In the simplest case, where the waveguide axis is along a $\langle 100 \rangle$ crystal axis, the nonlinear polarization direction coincides with the resonator axis, as described by:

$$P_z^{NL}(\Delta\omega) = d_{14}(\Delta\omega) E_{s1} E_{s2} \tag{8.3}$$

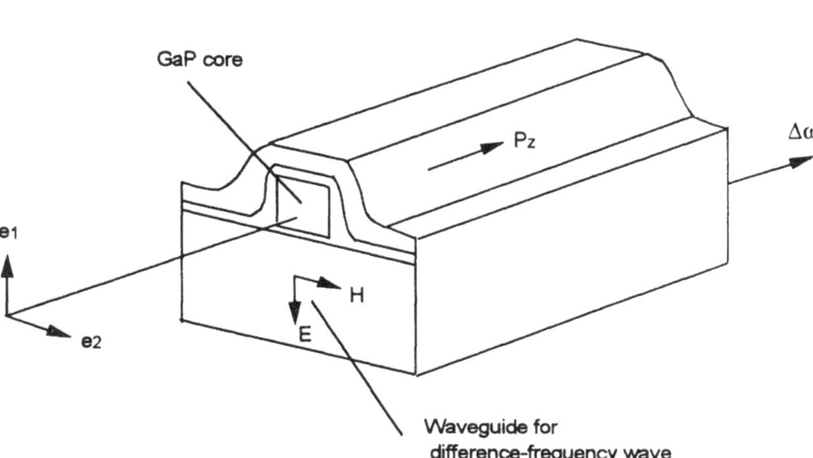

Figure 8.2 Frequency difference mixing by the semiconductor Raman laser. Incident pump beams ω_{L1} and ω_{L2} have polarization vectors $\mathbf{e}_1$ and $\mathbf{e}_2$ perpendicular to each other. Nonlinear polarization $P_z(\Delta\omega)$ is along the resonator axis. E and H are electromagnetic fields of the optical wave at frequency $\Delta\omega$ [3].

At the first sight, there seems to be no lightwave propagation at difference-frequency for the longitudinal polarization. However, it should be remembered that the light field propagating in a dielectric cylinder or a dielectric slab has a longitudinal component at the region near the surface of a dielectric waveguide. If the wavelength of the difference-frequency wave is much larger than the thickness of the Raman waveguide, the Raman waveguide can be disposed at the surface region of a waveguiding medium for the difference-frequency wave. However, if it is necessary to excite the transverse nonlinear polarization, the waveguide axis should be at a direction between a ⟨100⟩ and a ⟨110⟩ axis, although we do not go into details.

The output power at the difference-frequency, P_D, can be estimated by the following equation [4,5]:

$$P_D = T_D \left(\frac{\mu_0}{\varepsilon_0}\right)^{1/2} \frac{(2d_{eff})^2}{\pi n_1 n_2 n_p c^2} \omega_D^2 \frac{l^2}{A} \left(\frac{\sin^2\chi}{\chi^2}\right) P_1 P_2 \tag{8.4}$$

where

$$\omega_D = \omega_1 - \omega_2, \quad \chi = \frac{\pi l}{2l_c}, \quad l_c = \frac{\pi}{\Delta k},$$

A is the effective area of the incident beams, P_1 and P_2 are input optical powers, l is the interaction length, l_c is the coherence length, Δk is the momentum mismatching, T_D is the transmittance at ω_D, and d_{eff} is the effective electro-optic coefficient, which depends on the crystallographic direction. The estimated difference-frequency output power for this Raman oscillator can be as much as a few percent of the incident pump powers when the incident pump powers are $P_1' = P_2' \simeq 100$ mW, the waveguide cross section is 1 by 1 μm, the waveguide length is 1 cm, and $\omega_D \simeq 300$ cm^{-1}.

8.3 MATERIALS AND FABRICATION TECHNIQUES FOR THE RAMAN LASER

It was pointed out that most of the III–V compounds show very large values of the Raman polarizabilities compared to other classes of semiconductors. In a shorter wavelength up to a visible light wavelength, GaP is the most suitable. However at a longer wavelength region, a material with a narrower direct bandgap will be better because the Raman polarizability is larger as a result of the resonance enhancement. For example, at approximately 1.5-μm wavelength used for the long-distance optical communication, GaAs will be more suitable, although a GaAs—$Ga_{1-x}Al_xAs$ waveguide for the Raman laser will need much lower optical loss than that for a laser diode waveguide.

For further lower threshold operation, we should take into account the free-carrier absorption, deep-level absorption, and the two-photon absorption. The former two depend on the crystal quality, which will improve more and more by the control of epitaxy. As for the two-photon absorption, in our preliminary experiment, the two-photon absorption in GaAs was considerably smaller than that in InP at approximately 1 μm. Recently, the two-photon absorption in GaAs waveguides has been discussed [6].

The processes for waveguide formation raise problems about the perfectness of the waveguide [7]. The flatness of the waveguide walls must be further improved. At the same time, deep levels induced by the fabrication process must be reduced. One of the promising methods will be photo-etching. The practical length of the waveguide is only approximately 5 mm at the present. However, if the length can be made one order of magnitude larger, very convenient travelling wave light amplifiers will be available without the aid of regenerative amplification.

The superlattice structures have a possibility of enlarging the optical nonlinearities, including Raman polarizability, through the reduction of crystal symmetry and the occurrence of high strain fields.

REFERENCES

[1] C.H. Henry, "Theory of the Linewidth of Semiconductor Lasers," *IEEE J. Quantum Electron.*, Vol. QE18, 1982, pp. 259–264.

[2] J. Nishizawa and K. Suto, "Semiconductor Raman and Brillouin Lasers," *Infrared and Millimeter Waves*, Vol. 7, ed. by K.J. Button, New York: Academic Press, 1983, Chap. 6 pp. 301–320.

[3] K. Suto, *et al.*, "Semiconductor Raman Laser as a Tool for Wideband Optical Communications," *IEE PROC.*, Vol. 137, Pt. J, 1990, pp. 43–48.

[4] T. Yajima and K. Inoue, "Submillimeter-Wave Generation by Difference-Frequency Mixing of Ruby Laser Lines in ZnTe," *J. Quantum Electron.*, Vol. QE5, 1969, pp. 140–146.

[5] T.J. Bridges and A.R. Strnad, "Submillimeter Wave Generation by Difference-Frequency Mixing in GaAs," *Appl. Phys. Lett.*, Vol. 20, 1972, pp. 382–384.

[6] F.R. Laughton, J.H. Marsh and J.S. Roberts, "Intuitive Model to Include the Effect of Free Carrier Absorption in Calculating the Two-Photon Absorption Coefficient," *Appl. Phys. Lett.*, Vol. 60, 1992, pp. 166–168.

[7] K. Suto, T. Kimura and J. Nishizawa, "Semiconductor Raman Laser with Pump Light Wavelength in the 800 nm Region," *J. Electrochem. Soc.*, Vol. 140, 1993, pp. 1805–1808.

Index

The Artech House Optoelectronics Library

Brian Culshaw, Alan Rogers, and Henry Taylor, *Series Editors*

Acousto-Optic Signal Processing: Fundamentals and Applications, Pankaj Das

Amorphous and Microcrystalline Semiconductor Devices, Optoelectronic Devices, Jerzy Kanicki, editor

Bistabilities and Nonlinearities in Laser Diodes, Hitoshi Kawaguchi

Electro-Optical Systems Performance Modeling, Gary Waldman and John Wootton

The Fiber-Optic Gyroscope, Hervé Lefèvre

Field Theory of Acousto-Optic Signal Processing Devices, Craig Scott

Germanate Glasses: Structure, Spectroscopy, and Properties, Alfred Margaryan and Michael A. Piliavin

High-Power Optically Activated Solid-State Switches, Arye Rosen and Fred Zutavern, editors

Highly Coherent Semiconductor Lasers, Motoichi Ohtsu

Iddq Testing for CMOS VLSI, Rochit Rajsuman

Introduction to Electro-Optical Imaging and Tracking Systems, Khalil Seyrafi and S. A. Hovanessian

Introduction to Glass Integrated Optics, S. Iraj Najafi

Introduction to Radiometry and Photometry, William Ross McCluney

Optical Control of Microwave Devices, Rainee N. Simons

Optical Document Security, Rudolf L. van Renesse

Optical Fiber Amplifiers: Design and System Applications, Anders Bjarklev

Optical Fiber Sensors, Volume I: Principles and Components, John Dakin and Brian Culshaw, editors

Optical Fiber Sensors, Volume II: Systems and Applicatons, John Dakin and Brian Culshaw, editors

Optical Interconnection: Foundations and Applications, Christopher Tocci and H. John Caulfield

Optical Network Theory, Yitzhak Weissman

Optical Transmission for the Subscriber Loop, Norio Kashima

Principles of Modern Optical Systems, Volumes I and II, I. Andonovic and D. Uttamchandani, editors

Reliability and Degradation of LEDs and Semiconductor Lasers, Mitsuo Fukuda

Semiconductor Raman Lasers, Ken Suto and Jun-ichi Nishizawa

Semiconductors for Solar Cells, Hans Joachim Möller

Single-Mode Optical Fiber Measurements: Characterization and Sensing, Giovanni Cancellieri

For further information on these and other Artech House titles, contact:

Artech House
685 Canton Street
Norwood, MA 02062
617-769-9750
Fax: 617-769-6334
Telex: 951-659
email: artech@world.std.com

Artech House
Portland House, Stag Place
London SW1E 5XA England
+44 (0) 71-973-8077
Fax: +44 (0) 71-630-0166
Telex: 951-659
email: bookco@artech.demon.co.uk

www.ingramcontent.com/pod-product-compliance
Lightning Source LLC
Chambersburg PA
CBHW021430180326
41458CB00001B/206
9780890066676